THE CITY AS A TERMINAL

Transport and Mobility Series

Series Editors: Professor Brian Graham, Professor of Human Geography, University of Ulster, UK and Richard Knowles, Professor of Transport Geography, University of Salford, UK, on behalf of the Royal Geographical Society (with the Institute of British Geographers) Transport Geography Research Group (TGRG).

The inception of this series marks a major resurgence of geographical research into transport and mobility. Reflecting the dynamic relationships between socio-spatial behaviour and change, it acts as a forum for cutting-edge research into transport and mobility, and for innovative and decisive debates on the formulation and repercussions of transport policy making.

Also in the series

Ports, Cities and Global Supply Chains
Edited by James Wang, Daniel Olivier, Theo Notteboom and Brian Slack
ISBN 978 0 7546 7054 4

Achieving Sustainable Mobility
Everyday and Leisure-time Travel in the EU
Erling Holden
ISBN 978 0 7546 4941 0

Policy Analysis of Transport Networks
Edited by Marina Van Geenhuizen, Aura Reggiani, and Piet Rietveld
ISBN 978 0 7546 4547 4

A Mobile Century?
Changes in Everyday Mobility in Britain in the Twentieth Century
Colin G. Pooley, Jean Turnbull and Mags Adams
ISBN 978 0 7546 4181 0

Rethinking Urban Transport After Modernism
Lessons from South Africa
David Dewar and Fabio Todeschini
ISBN 978 0 7546 4169 8

The City as a Terminal
The Urban Context of Logistics and Freight Transport

MARKUS HESSE
University of Luxembourg

Routledge
Taylor & Francis Group

LONDON AND NEW YORK

First published 2008 by Ashgate Publishing

2 Park Square, Milton Park, Abingdon, Oxon OX14 4RN
711 Third Avenue, New York, NY 10017, USA

Routledge is an imprint of the Taylor & Francis Group, an informa business

First issued in paperback 2016

British Library Cataloguing in Publication Data
Hesse, Markus
  The city as a terminal : the urban context of logistics and
  freight transport. – (Transport and mobility series)
  1. Terminals (Transportation) 2. Urban economics 3. City
  planning 4. Freight and freightage
  I. Title
  388.3"3

Library of Congress Cataloging-in-Publication Data
Hesse, Markus.
  The city as a terminal : the urban context of logistics and freight transport / by Markus
Hesse.
    p. cm. -- (Transport and mobility)
  Includes bibliographical references and index.
  ISBN 978-0-7546-0913-1
  1. Delivery of goods. 2. Urban transportation. I. Title.

  HF5761.H48 2008
  388".044--dc22

                                                                                      2008025990

ISBN 13: 978-0-7546-0913-1 (hbk)
ISBN 13: 978-1-138-25520-3 (pbk)

Contents

List of Figures

List of Maps

List of Tables

Preface

The contents of this book originate from several years of research on the relationship between city and region on one hand and logistics and freight distribution on the other hand. This subject remained somehow hidden for a considerable amount of time, since spatial science in general and urban, economic or transport geography in particular did not investigate logistics changes and their spatial outcomes, except a handful of studies that had been conducted either in the 1960s or more recently in the 1980s and 90s.

The starting point of the research was the observation that many of the developments and problems associated with what was then called "urban logistics" – such as congestion and truck traffic, noise emissions and air pollution, land use and community issues – had been shifted toward more remote places, particularly toward suburban areas, as a matter of corporate locational behaviour (Hesse 1998). As a consequence places which were formerly dedicated to the transshipment and handling of goods, namely city ports and warehousing districts, became transformed and redeveloped to urban waterfronts or loft houses. In contrast to the rising awareness that these urban transformations had received, only little attention has been paid to the locales where goods distribution had moved to: sub- or even ex-urban areas with excellent transport access and apparently unlimited supply of cheap land. Paradoxically, it was the advent of the Internet and the often exaggerated emphasis on the immaterial, volatile cyberspace that had subsequently re-shifted the focus on material flows and places.

Meanwhile, a rising interest has been directed towards the issue of mobility and flows in geography (cf. Hall et al. 2006). This is due to new corporate strategies in manufacturing and distribution, the application of new technologies, also an increased competition and the challenge of globalization at all. As a consequence, new logistics networks emerged, consisting of nodes, hubs and flows in a broad range of interaction. Thus logistics and freight distribution became the transmission belt for global trade and expanded production systems. In this context, places that function as the interface between sender and receiver, shipper and customer have attracted special attention: port areas, export processing zones, logistics parks that host distribution centres, dedicated freight rail yards, commercial and industrial areas with wholesale trade, retail distribution and freight forwarding, just to name a few but important examples.

Against this background, the particular interest of this study was to investigate the emergence, dynamics and character of some of these places. As a research subject, they are considered being "hybrid" – a compositum of industry and services, of derived and intrinsic activity, a driver and a product of industrial structural change, and finally an outcome of both the global and the local. As far as the study tied up to earlier inquiries into the characteristics and conflicts of urban freight distribution, it was particularly concerned about corporate decisions for leaving urban areas and relocating elsewhere. The study also dealt with the rationale for other firms to remain in the core city. Moreover, the shift of distribution firms to the urban periphery or even beyond has been assessed in relation to earlier processes of commercial and industrial suburbanization, looking at the potential consequences for both the city and the suburbs. The findings reveal in more detail how firms are assessing these different and often contradictory advantages and disadvantages of urban places, in order to identify the optimal, or at least "second best" mode of operation and the appropriate location. Consequently, the study is also about how public agencies, such as municipalities or regional development agencies, are dealing with logistics and freight distribution, both as a problem and as a potential.

This study is geographical by nature, since it attempts to shed light on the relationship between fixity and motion, between place and mobility. Places that are dedicated to organize flows can actually be considered a contradiction by definition. Speaking in terms of David Harvey (1989), the commitment of logistics to optimize time is associated with a certain "annihilation of space". In so doing, logistics constantly works at reinforcing the de-territorialization of the economy, since it allows for not only to move commodities but to re-position economic activity. However, in contrast to other fields of economic geography research, there are no vital signs indicating that a certain re-territorialization of this sector is underway, as it is the case e.g. with the knowledge economy and the related networks or clusters. Yet, even the emerging greenfield investments in container ports, "freight villages" or "cargo cities" (sic!) at airports and in newly developing off-shore facilities may depend upon establishing a new territorial context. Though logistics is primarily not about place, it is about organizing flows as efficient as possible, thus enabling places to participate in the increasingly connected economy. Consequently, it is of major interest for the discipline to learn more about the territorial arrangements and entities that are resulting from the particular interplay of fixity and motion that is embodied by logistics changes. Keeping in mind the rising interest in flows and mobility expressed by social sciences in general and human geography in particular (Crang 2002), there are significant issues to be addressed.

Acknowledgements

As any major research publication, the volume presented here has substantially benefited from the input provided by institutions and people who made my research possible or who challenged my very early hypothesis and helped putting it the right way. First and foremost, major support was provided through a post-doctoral grant provided by the German Research Organization (Deutsche Forschungsgemeinschaft, DFG), which lasted from 2000 to 2003 and made this study essentially possible. It was complemented by several conference travel grants offered by the DFG that allowed for establishing and maintaining fruitful intellectual exchange with colleagues from abroad, particularly with Jean-Paul Rodrigue from Hofstra-University, New York (USA), Peter V. Hall, Simon Fraser-University, Vancouver (Canada) and Theo Notteboom from the University of Antwerp (Belgium).

I am also indebted to Professor Gerhard O. Braun of the Urban Studies Unit of the Freie Universität Berlin, who gave my research every possible scientific, material and organizational support from the very beginning. It was mainly due to his persistent encouragement that a previous version of this study was successfully submitted and accepted as a "Habilitationsschrift" by the Faculty of Earth Sciences of the Freie Universität Berlin in February 2004. Particular credits go to Professor Karl-Dieter Keim, Berlin, once the Director of the Institute of Regional Development and Structural Planning (IRS) in Erkner (near Berlin), who was very supportive even before I officially joined the Institute in 1998, and thus made my subsequent research leave possible at all. More than 130 people were available for expert interviews in the context of the two case studies that were carried out between 2000 and 2003 in the Berlin-Brandenburg region and in the San Francisco Bay Area. They deserve a particular consideration. Betty Deakin and Marty Wachs at the University of California at Berkeley (USA) were extremely helpful in making my time as a Visiting Scholar at UC Berkeley's Transportation Center as comfortable and successful as possible.

Several people were instructive once this manuscript and its original predecessor on "Goods transport and logistics in the process of urbanization" had been put together, notably Sabine Meister (cartography), Britta Trostorff and particularly Steven Bayer for data processing, GIS and related activities that helped making the text presentable. Anne Beck from the Faculty of Geosciences of the Freie Universität Berlin was very supportive by providing English translations of chapters, also Diana Hömann who gave the manuscript a careful proof-reading.

My final thanks go to Ashgate publishers, particularly to Carolyn Court and Val Rose for support, post-production, and, even more important, patience with serious delays in submitting the manuscript. Special thanks also go to Richard D. Knowles from the University of Salford (United Kingdom), and Brian Graham from the University of Ulster (Northern Ireland), who accepted to include the volume in the Ashgate series on "Transport and Mobility" as series Editors, but who also had to wait for my output far too long.

References

Crang, M. (2002), "Commentary", *Environment and Planning A* 34, 569–74.

Hall, P., Hesse, M. and Rodrigue, J.-P. (2006), "Re-exploring the interface between transport geography and economic geography". *Environment and Planning* A 38:8, 1401–08.

Harvey, D. (1989), *The Condition of Postmodernity*. (Oxford: Blackwell).

Hesse, M. (1998), *Wirtschaftsverkehr, Stadtentwicklung und politische Regulierung. Zur Bedeutung des Strukturwandels in der Distributionslogistik für die Stadtplanung*. (Berlin: Deutsches Institut für Urbanistik) = Beiträge zur Stadtforschung 26.

Introduction

The City as a Terminal.
Logistics and Freight Distribution in an
Urban Context

Background and Rationale of the Study

The exchange of goods is a constant feature of human economic activity. It was once essential for the rise of the mercantile economy in medieval Europe (Braudel 1982) and became a large scale activity during the industrial revolution. The location of industry and thus the geography of manufacturing in general evolved with respect to accessibility improvements that were particularly offered first by ocean shipping and inland waterways, later by railways which were then predominantly freight related. Vice versa, every "long wave" in the process of industrialization embodies distinct transport orientations and appropriate infrastructure requirements (Hayter 1997, 27). This was true for the railroad in the Fordist economy, as it is for the trucking and air freight modes more recently. Even the modern information and communications technologies may currently represent just another step in this incremental co-evolution of economic development and physical distribution, rather than to disrupt with earlier stages and to bring about totally new modes of value creation.

The spatial organization of economic activity has been fundamentally transformed over recent decades, as an outcome of structural changes, new technologies and particularly globalization: the expansion of world trade, manufacturing and goods distribution across the globe. It is often overlooked that the rising exchange among different parts of the world is associated, first, with logistics organization, comprising the management of the supply of raw materials and components, goods manufacturing, and the physical distribution of the product to its final destination. Logistics aim at delivering consignments in the right composition (in terms of quantity and quality), at the precise time and lowest possible cost, which also equals a formal definition. Second, it requires physical activity and infrastructure, particularly the transfer of commodity shipments by truck, freight rail, airplane, waterway or ocean ship, the handling of consignments in warehouses, distribution centres (DC) and parcel stations, and the delivery of shipments to the final point of consumption. The material dimension of logistics and freight distribution remains prevalent, despite the emergence of electronic commerce and web-based management practices, promoting the image of apparently virtual flows that need no material infrastructure and physical distribution.

However, recent innovations in technology and management have in fact led to significant changes in logistics organization and operation. This applies particularly to the integrated, upstream management of supply chains and the re-organization of distribution networks, with a much higher spatial reach than before. As a result of these transformations, logistics has changed radically since the post-war decades. It may no longer represent just a derived function which is primarily dependent on the demand for services by manufacturing or retail firms, but also a powerful system that lays the ground for a flexible, highly distributive economy. It increasingly follows a distinct logic, likely to influence related parts of manufacturing or retail, rather than being determined by the place and the time of production or distribution.

The extent to which logistics is changing the landscape is being sketched by *The Economist* as follows:

> Historically, transport technology has always made a physical impact on centres of commerce. In the days when cargo was loaded onto ships mainly by hand, factories would often cluster nearby because transport costs were high and delivery slow. With the arrival of the shipping container, factories were able to move to cheaper locations and away from crowded city ports such as New York City and the London Docks. Container terminals did not have to be so close to large population centres, provided they had plenty of space, railways, goods roads and workers prepared to handle containers, which many stevedores in older ports were not. Somethin similar is now happening around logistics centres, especially at airports. Companies are moving some or all of their operations to be near such centres because this allows them to process orders late into the day and put their goods on the last flight out, for delivery the following morning. [...] The idea that if you build a logistics centre companies will come is being taken to extremes in the United Arab Emirates. [...] Now Dubai is building what it describes as the world's first "logistics city". (*The Economist*, 17th June 2006, 13)

There is no doubt that the spatio-temporal extension of goods movement has important implications for urban and regional development. Access to markets, linking supply of and demand for goods, being a "central place" for city and hinterland as well as for distant trade have played a major role in the emergence of the city, both in mercantile and industrial ages. Large technical systems such as logistics and freight distribution are embedded in a particular spatio-temporal framework: they require certain material facilities, infrastructure and "space for operation", thus re-arranging the conditions of space and time in a broader context. Sufficient logistics capacity determines accessibility and influences the economic prospect of certain places, so the precise functionality of delivery is still an important factor of the urban and regional economy – even if these places are under increasing pressure of cost reduction and on-time functionality.

Despite this basic significance of logistics and freight distribution, neither geographical studies nor transport research paid any particular attention to this subject until recently (see the overview in chapter three). Spatial studies still lack a considerable understanding of logistics organization and freight distribution, which particularly applies to the role that cities and urban development play in this respect. In turn, the relationships between logistics and spatial or urban development are widely neglected by business management and freight transport planning, both being the traditional disciplines of logistics and freight-related investigation. Even the emerging study of global commodity chains pays little if no attention to the underlying framework of physical distribution and the respective role of infrastructure, connectivity or transport costs. Judging from current geographical research, logistics and freight distribution remain a "missing link": between the widely investigated new production systems and the outstanding work on (post-) modern consumption patterns. This study is an attempt to link these fields together and to illuminate the fundamental role logistics and freight distribution play for urban and regional development. Moreover, the study tries to prove how logistics and freight distribution are influenced by spatial structure and variation.

Main Hypothesis and Theoretical Angles

The main statement of this book is that modern logistics is shaping urban development and urban land use, as a consequence of new supply chain organization and logistics network design. This transformation of urban places includes, first, the re-development of warehousing districts, inner-city rail yards and freight consolidation facilities, in favour of more valuable and competitive land uses, such as housing, retail or business services. This phenomenon also applies to the increasingly popular conversion of city ports into urban waterfronts. The underlying rationale is based on changing preferences by logistics and freight distribution firms that had disadvantaged core urban areas, particularly the demand for "big box" space and the imperative of high-throughput distribution in 24/7 modes. Both properties can no longer be offered by inner-city locations, since both land rents and sensitive land uses do not permit the ongoing spatio-temporal extension of logistics and freight distribution.

Second, as a consequence, facilities that host logistics services are increasingly being re-located toward strategic places within and beyond urbanized territory. This applies particularly to functions such as storage, consolidation and high-throughput distribution of consignments. The related structural changes are creating new geographies of distribution, as an outcome of supply chain re-organization and logistics network design. They can be observed at all spatial levels, e.g. in port regions, which are the major nodes of the global supply chains, yet also in ordinary agglomerations, where land is scarce and both agglomeration disadvantages and regulations are driving distribution land uses towards the periphery.

The resulting shift out of urbanized areas toward sub- or ex-urban places has already been labelled as "port regionalization" in the case of port regions, or as "logistics polarization" in more generic terms. These movements imply more than just spatial shifts, as they are shaping the function and the character of urban places: They affect both the traditional role of the city as a centre of goods merchandising, which is becoming re-designed under the flag of globalized distribution regimes, as well as urban structure and urban land use. Whereas both spatial tendencies, the move to the suburbs and the preference for remaining in core city areas, can be confirmed as a regular locational practice of distribution firms, the former appears to be the predominant pattern. This is particularly the case since logistics is increasingly performed by major corporations that operate large-scale networks, with expanded spatial reach and particular emphasis on mobilizing economies of scale. This business model can hardly be practised in core urban areas, as this requires extensive space reserves and unimpeded traffic conditions.

Accordlingly, the book investigates two different spatial dimension of logistics and freight distribution: it discusses the generic urban attachment of logistics functions and related processes of dissociation from the city, and it also focuses on urban-regional locational dynamics and conflicts, with particular emphasis on suburban areas as the major logistics "organization space" which emerged due to changing organizational settings and locational requirements. In the latter section of the book, planning and political regulation of such land uses are discussed as well, since the steadily expanding system of physical distribution exterts a significant pressure on urban areas, on neighbourhoods adjacent to freight sites and on the transport system. The related theoretical background of the study is based on a combination of different approaches from economic and urban geography, reflecting the hybrid character of the subject, and it aims to illuminate the interrelationship of three basic processes:

- the modernization of corporate logistics (by shippers, freight forwarders, courier and parcel services and also wholesalers) in the context of a flexible production and distribution organization;
- the determination of location-choices of such companies according to logistical, transport and space requirements;
- the urban development with respect to the distribution function and particularly the contribution of logistics to the dispersal of the urban region.

The related multidimensional research concept draws upon a trans-disciplinary perspective, including the following three angles:

First, against the background of sectoral shift and technological change, the study discusses the post-Fordist economy and flexible specialization which are emblematic for late-industrial change, particularly the establishment of decentralized production networks (Dicken and Thrift 1992). These production systems imply, if not dissolution then a re-arrangement of the locational fixity of the single firm. Logistics and freight distribution are an inherent part of the resulting large scale value-added chains and networks in the sense that they enable economic actors to organize these chains and networks at a great distance.

Second, urban de-concentration, suburbanization and the emergence of the poly-centric city-region build the appropriate spatial framework at the regional level (Walker and Lewis 2001; Kloosterman and Musterd 2001). The locational dynamics of logistics and freight distribution firms (wholesale, logistics, freight forwarding, shipping) will be assessed in this specific context, which also includes the role of cities in widely stretched logistics chains and distribution networks.

Third, an institutional perspective on urban and regional development (Amin 1999) is chosen in order to highlight the particular inter-relation and interaction of corporate demand for location, the supply of land and the role of intermediaries such as land- use planning and economic development.

The advantage of such an approach is to overcome traditional assumptions on corporate locational behaviour, mostly based on a catalogue of location factors that attract or firms to settle at certain places (Hayter 1997, 5, 6). Empirical evidence suggests, first, that regional development processes are more complex, locally embedded, and context dependent than assumed in conventional location theory (Storper 1997, 26, 39), and, second, that infrastructure provision, private investments, policy and planning are mingling in mechanisms of cumulative causation, creating inter-related processes of local growth. These processes are interpreted in a special way: that as a consequence of corporate location choice and further development, places are going to become "produced" (Storper and Walker 1989, 70). The starting point of this approach is a cyclical understanding of regional development with a periodical change between de-concentration and re-agglomeration (*localization, clustering, dispersal, shifts*). This model basically assumes that the environment of regional development is unstable, dynamic and that development mainly depends on the conditions and carriers of growth (Storper and Walker 1989, 9). The challenge arising for empirical inquiry is to assess the role logistics and freight distribution play in this particular context of "geographical industrialization" or "industrial dispersal" (ibid.). Given the particular demand of locations dedicated to logistics and freight distribution for cheap space and extraordinary traffic accessibility, these investments might even work as "pioneers" in developing and thus "industrializing" certain areas.

Definition of the Research Subject

Logistics consider a wide set of activities dedicated to the transformation and circulation of goods, such as the material supply of production, the core distribution and transport function, wholesale and retail and also the provision of households with consumer goods as well as the related information flows (Handfield and Nichols 1999). These activities composing logistics are included into two major functions which are physical distribution, the derived transport segment; and materials management, the induced transport segment. These are also subjects that will be investigated in this study in more detail.

Physical distribution is the collective term for the range of activities involved in the movement of goods from the points of production to the final points of sale and consumption (McKinnon 1988, 133). It comprises all the functions of movement and handling of goods, particularly transportation services (trucking, freight rail, air freight, inland waterways, marine shipping, and pipelines), transshipment and warehousing services (e.g. consignment, storage, inventory management), trade, wholesale and, in principle, retail. Conventionally, all these activities are assumed to be derived from materials management demands. Physical distribution must insure that the mobility requirements of supply chains are entirely met.

Materials management considers all the activities related to the manufacturing of commodities in any stage of production along a supply chain. Materials management includes production and marketing activities such as production planning, demand forecasting, purchasing and inventory management. It must ensure that the requirements of supply chains are met by dealing with a wide array of parts for assembly and raw materials, including packaging (for transport and retailing) and, ultimately, recycling discarded commodities. All these activities are assumed to be inducing the demand for the aforementioned physical distribution.

The close integration of physical distribution and materials management through logistics is blurring the traditional distinction of induced and derived demand for the related services. The reciprocal relationship between the induced transport demand function of physical distribution and the derived demand function of materials management is thus considered as the integrated transport demand of logistics. This implies that distribution, as always, is derived from materials management activities (namely production), but also, that these activities are coordinated within distribution capabilities. Production, distribution and consumption are therefore difficult to separate.

In this context, the subject of inquiry is defined as those subsections that are 1) statistically distinct from other subsections (e.g. marketing in firms, or passenger modes in transport), 2) that are locationally mobile and not confined to the network of inland waterways or railways. Accordingly, the respective statistical units researched in this study are freight transport, warehousing, and freight forwarding (including freight transport arrangement), also couriers and messengers.

The Urban Context

Logistics and freight distribution are considered to be the key issues of urban and regional development. In this respect, the book ties up to a traditional characteristic of the city: serving as a major node for the creation of commodities or at least for their exchange and delivery. Cities have always been the predominant places of goods handling and logistics services (Hesse 1998a). Under the current framework conditions, the book concentrates on the urban field for two major analytical reasons: Cities are sources and destinations of a majority of the freight flows, and speaking in historical terms, many of them have emerged due to the trade function and as a market place respectively (see Chapter 1).

More recently, and due to logistics modernization against the background of globalization and technological change, cities are no longer considered the prime locale of economic activity (Amin and Thrift 2002). This applies predominantly to the industry that underwent a significant de-centralization over the past decades, yet may also have occurred in logistics and freight distribution, indicating a certain de-linking of logistics and the urban. However, there is a constant demand for goods delivery generated by and in urban areas as long as city regions host a large portion of consumers. Also, the diversity and sensitivity of urban land uses is confronted with the somehow "harsh" nature of truck and rail operations. This causes many conflicts between freight distribution and other urban functions. Freight traffic operated by heavy goods vehicles represents a significant burden for the natural and the social environment of cityscapes. According to OECD research, the freight segment is likely to contribute to approximately 30 per cent of transport-related energy consumption (which accounts for about 20 per cent of all energy consumption in advanced economies) – so there is little doubt that dealing with logistics and freight distribution is important.

Despite an increasing awareness for both the functionality and sustainability of freight distribution and logistics, local attempts to manage urban freight transport have not yet provided sufficient answers (OECD 2003). This is primarily due to the complexity of the chain, since it bundles a broad range of units, actors and interests. Cities also tend to be situated at the end of the logistics chain without any major influence on its management. The main factors behind these practical problems are the intermediate character of physical distribution within network economies (a feature that might be similar to the role of mobility in network societies), and the permanent but diffuse reproduction of freight transport demand and supply through time-space bound interaction. On the one hand, the low number of successful examples of planning intervention, not to mention good or best practice, is indicative. On the other hand, urban areas and the related planning regime represent an appropriate framework for the governance of logistics and freight distribution, since urban densities generate a basic need for spatial and temporal adjustments of goods flow to the city.

Study Design, Methodology and Research Questions

Empirical evidence on the socio-economic and spatial dimensions of logistics restructuring is provided by two comparative case studies. The methodological rationale for this is based on the premise that this complex field of interaction, which has hardly been investigated until now, cannot be accessed through the generation and assessment of macro-data, e.g. on materials flows or traffic counts. First, it is necessary to identify the role that logistics and freight distribution play in the specific urban context. Second, against this background, the interplay of functional and spatial logics of the system of flows and the space of places, has to be conceptualized. Third, corporate decisions on where to place logistics facilities have to be reconstructed in order to understand the particular properties of different places to attract or to deter corporate investment. This multifaceted character of the subject has been investigated by conducting expert interviews in firms, the real estate business and with urban planners.

One of the two case studies was carried out in the Berlin-Brandenburg Metropolitan Area, Germany, between 2000 and 2003, the other in the San Francisco East Bay Area/ Central Valley in Northern California, USA, between 2001 and 2002. Both regions are witnessing a significant degree of logistics re-organization, associated with spatial restructuring at the metropolitan level. The German region is subject to post-socialist transformation, with de-industrialization and an emerging service sector, creating what can be called a post-industrial distributive economy. By contrast, the North American region represents a new industrial space based on high-technology manufacturing, network building and innovation. These apparently different cases represent a broad range of characteristics of both industrial and post-industrial societies and the way these are becoming reproduced in space and time. Despite some clear differences in terms of regional framework conditions between the two cases, there are serious commonalities regarding corporate decision making and the way locational attributes and logistical arrangements are evaluated and refined in order to achieve the best possible organization of distribution. The results of the case studies are assessed against the background of large-scale developments in both countries, namely supply-chain integration, port and DC-regionalization, and the emergence of new technologies and their impact on logistics and freight distribution (for an overview of the research concept see Figure 0.1).

In order to reconstruct the locational dynamics of logistics and freight distribution firms, it is essential to track their spatial behaviour, based on statistically available indicators. In the following, particularly in the two case studies (chapters four and five), the subject of logistics and freight distribution is investigated with particular respect to locational dynamics in urban regions. Particularly in the case studies, the related subsectors have been investigated with respect to four major sets of research questions:

- Where do physical distribution firms currently locate? What are the major factors that determine the locational decisions of distribution firms in these areas? Have there been changes in the locational pattern over the last few years, following the assumption that there is a general movement of distribution facilities from core areas towards the urban fringe?
- Is there a particular contribution of logistics firms triggering sub-urbanization or even ex-urbanization, e.g. by the initial movement of "pioneer" firms out of town, thereby opening the door for others to subsequently follow?

- How, if yes, do political authorities include the freight sector into their planning strategies, and how do they cope with the costs and benefits delivered by logistics?
- What do the findings mean for assessing the economics of the contemporary city, i.e. with regard to the future role of physical distribution in an urban context?

Two basic methods have been applied in an attempt to answer these questions. Statistical analysis have been carried out, primarily based on the assessment of employment data from census sources, at county and city levels, for the goods distribution industry, but also of warehousing inventory data retrieved from real estate firms. Selected planning and economic development sources and documents provide an additional source of data. Location quotient analysis at county and city level is used to identify concentrations of the distribution industry relative to, or different from, the changes in other economic sectors. The employment data for Germany were provided by the Federal Office of Employment (*Bundesagentur für Arbeit*) and are related to the county and municipality levels. In the U.S., the data originated from the County Business Patterns, delivered by the *U.S. Census*, and solely refers to the county level. In both cases, certain time series reveal historical records; however, there are some limitations for data interpretations due to changing data classifications.

A qualitative survey had been conducted among relevant agents in the physical distribution industry. In Berlin-Brandenburg, 23 regional experts were interviewed in autumn 2000, followed by a corporate survey of a project with 50 firms in 2002 and 2003. In the East Bay Area and the Central Valley, about 50 personal interviews had been conducted between June and September 2001. Among the surveyed people were: first, city, county and state officials in the economic development, transportation, and land use sectors; secondly, corporate executives

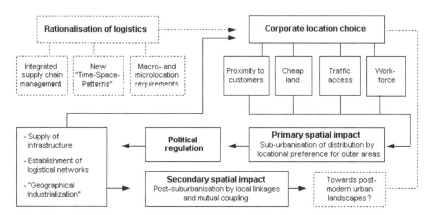

Figure 0.1   Overview of the research

Source: Author's own.

in distribution firms such as transportation and warehousing, retail distributors, freight forwarders and carriers, container shippers, port and airport-related experts; thirdly, real estate agents who are active in industrial and commercial real estate transactions were interviewed. In addition to that, 20 personal interviews had been conducted with researchers in the field of Geography, City and Regional Planning, Logistics, Transportation, and Architecture in order to cover a range of different disciplinary perspectives.

It is inevitable that both case study areas are shaped by singular circumstances, or region-specific features of transformation. But the results that have been produced may go beyond the two case studies to identify general lessons or principles of change. This problem of separating the general from the particular can be answered in two ways. Transatlantic comparisons are (being) carried out from theory and empirical findings regarding the similarities and differences between the case study regions and their socio-economic framework conditions. Yet the problem of providing generalizations from only two case study areas is recognized. Hence for comparative reasons, a literature review had been conducted in order to systematically relate the two angles of the research to each other: logistics and freight distribution on one hand, and their locational dynamics in the context of urban development and suburbanization on the other hand. Related developments that have been observed in other regions recently, such as Hamburg and the Ruhr Area in Germany, Benelux in western Europe and also in New York-New Jersey and the Greater Los Angeles area in the U.S., have been set in relation to the results of the review.

The Structure of the Book

The starting point of the book is an introduction that gives an overview of the main fields of research and the related questions that have been investigated in the underlying inquiry. Research hypothesis are developed in order to prove and answer these questions. The introduction is followed by a theoretical and literature-based discussion of the two major components of the study context: logistics and freight distribution in an economic geography context on one hand and urban development and related characteristics of the city on the other hand. Therefore, classical and contemporary geographical models and developments are discussed in their general significance for the research as well as contemporary logistics development is assessed in terms of spatial implications. Thus Chapters 1 and 2 of the book are dedicated to conceptualizing the subject with respect to the related frameworks. The third chapter ties together these two dimensions and aims at putting the subject of inquiry in a particular geographical context, which is also coined "geographies of distribution". Chapters 4 and 5 present the findings of the two case studies in a condensed but detailed manner where necessary. The case studies are supplemented by a short account of locational dynamics of logistics and freight distribution in terms of policy and planning which constitutes

Chapter 6. There are certain activities reported at the regional level that try to cope with both the desired economic impact of logistics and the negative externalities of freight transport. Chapter 7 bundles the main findings of the study in general, both with respect to the case study regions as well as in terms of generic lessons to be learned, and reformulates the significance of logistics and freight distribution for urban development even under late-modern conditions.

Chapter 1
The City – From Market Place to Terminal

Cities as Markets: Trade and Transport as Origins of Urban Places

Cities and urban regions have always been significant nodes for the exchange of goods. Trade and merchandising, wholesale and retail distribution have been closely connected with urban places and urban development. Cities have been a central place by definition, for both city and region, and a gateway for providing goods and services to a more distant hinterland. Not coincidentally, the classical function of the city as a centre of goods transshipment has already been acknowledged in traditional urban theory. The sociologist Max Weber (1921, 61), for example, argued that the "regular exchange of goods" was one of the basic characteristics of cities, distinguishing a city from other, more or less ordinary settlements. One of the classical studies in urban and regional development, Walter Christaller's Central Place Theory (Christaller 1933), had put particular emphasis on 1) the significance of the city for providing goods and services for the city and for the area beyond, 2) on the role of transport costs for defining its sphere of influence both from the perspective of the supply side as well as from the consumers point of view. The ability of the city to concentrate people and workforce, ideas and interactions, can be considered a constant factor in the process of urbanization. The spatial manifestation of this force of concentration was the significant role of the urban centre that lasted for centuries – until new means and technologies of mobility were changing the shape and the structure of the city. With respect to these changes, the theory of the multi-nuclear city developed by Chauncy D. Harris and Edward L. Ullman (Harris and Ullman 1945, 9–15) emphasized the foundation of the city as a generic transport focus, emerging from the demand for "break of bulk" and related services and amenities. This generic urban significance of the system of goods exchange was particularly fostered by interregional trade that made some cities becoming nodes within a large-scale network of commodity and money flows. Once analysing the emergence of the medieval trade economy, the French historian Fernand Braudel described urban places that were committed to organize trade and retail as "trading spaces" (Braudel 1982, 26 ff.), like market places or market halls, warehouses and finally trade fairs. Port cities performed and further developed such properties in a particular way. Some of these cities, like the port city of Hamburg in Germany, the old trade city of Antwerp in Belgium or the prototypical "gateway city" in Chicago in the U.S., have retained this property until now. Others developed in different directions and have specialized in manufacturing or services, more recently in high-technology, knowledge-intensive activities or leisure and tourism. Even in these cases, providing accessibility for the flows of goods and people (and also information), remains important for urban development in general.

Theories on urban structure and urban land use had also shed some light on the importance of trade and the related movements. Burgess' classical study on the concentric model of the city of the twentieth century included the emergence of specific districts for hosting the provision of commodities, such as wholesaling or light manufacturing in the zone of transition, which was traditionally located close to the city centre. Since manufacturing was once concentrated in the urban core, a major reason for the location of industry and derived from that of distribution was explained by the advantages of urban agglomeration in terms of transportation and labour orientation, according to Alfred Weber's theory of the location of industries (Weber 1929, 41 ff., 95 ff). As a consequence, the provision of goods was important for the development of particular urban places confined to storage, goods handling and processing. Such places were found both on-site of the industrial plant and adjacent, also in dedicated commercial areas or in more remote facilities, respectively (see Table 1.1). The remainders of this period are so-called wholesale districts with warehouses and loft-buildings, also urban waterfront areas with past and present transshipment functions, or secondary industrial warehouses. Due to changes in urban land use, and in response to new logistical requirements, these places have been the subject of urban regeneration projects and are normally no longer associated with organizing material flows (Hesse 1999).

Table 1.1   The traditional place of goods handling in urban areas

	Function	Location	Examples
The city as a market place	Traditional place of goods exchange. (The city as a location for regional distribution)	Historical urban centres; temporary use of areas for warehousing and transshipment	Market places, traditional locations for urban retail, warehouses
Port cities, Inland-Port cities	Traditional place of goods exchange. (The city as a location for long-distance distribution)	Traditionally at shorelines, large inland waterways, intersections of distant trade-routes	Ports and port-infrastructures, storage buildings, warehouses, magazines
Rail freight terminals	Development of new transshipment points according to the industrial urbanization	Main stations and their backyards (close to the urban core, e.g. in "zones of transition")	Rail terminals and railyards, until recently in all major cities with railway access
Wholesale, Freight Forwarding	Suburbanization of distribution functions out of the core city (first outward drift)	Urban peripheric locations, close to highway intersections	Transportation intensive land uses (commercial, industrial areas)

Source: Compiled by author

The Industrial Revolution and the accelerated urbanization that occurred during the last third of the nineteenth and the beginning of the twentieth century have underpinned these developments. Industrial urbanization can not only be understood as a special mode of creation of value through materials transformation, but it became strongly connected with the development of mass transport technologies in general and the railway system in particular. The emerging "sphere of circulation" was of course extremely important for the shape and the structure of the industrial city. Once rail freight and passenger transport became predominant modes of distribution for decades, a new network infrastructure had been exposed to urban and suburban areas providing for accessibility in a new format, also subdividing and fragmenting urban space due to the specific layout of the roadbed. Beyond the mere establishment of rail tracks and stations, transshipment terminals and extensive railyards were established that covered a significant amount of space in urban core and fringe areas. Altogether with large manufacturing plants, this setting would later be considered a typical image of the "Fordist" city. Again, the making of the gateway city of Chicago (Cronon 1991) had been strongly associated with a new centrality and an expanded spatial reach offered by the railway system.

The coming of the motor truck has altered this picture again. This happened in the U.S. during the first half of the twentieth century, in most industrialized countries such as in Europe after World War II. The direct door-to-door services offered by the motor truck allowed for a completely new flexibility in terms of space and time – a seemingly appropriate response to the increasingly dispersed urban landscape and fragmented spatial economy. In one of the rare contributions that covered the influence of goods movements on urban and suburban areas, notably that of trucks, Jackson (1992, 1994) pointed out as follows:

> Truck transportation in the 1920s had proven already that it was fast, flexible and cheap, so one important reason for wanting a new layout for factories and warehouses was the provision of efficient loading and unloading facilities for the truck – facilities that integrated the doors and tailgate of the truck into the horizontal interior of the plant. [...] Accordingly, factories and warehouses began deserting the crowded town area and moving out to where land was cheaper and closer to the highways used by larger trucks. So the truck – or the increased use of the truck – contributed to the decay of the inner city and the growth of industries in the outlying districts. (Jackson 1992, 21)

So the motor truck forced the drift of economic activity away from the fixed topography determined by the location of cities and the layout of railway lines (as the refrigerator wagon had triggered the massive expansion of economic catchment areas earlier). These logistical innovations reinforced the growth of freight transport, and they stimulated urban and regional development in general. The main comparative advantage of the motor truck compared to rail and waterway transport was to combine spatial reach with temporal flexibility; moreover, the truck made the change of vessels that had happened before in ports or railyards mostly obsolete, thus reducing delivery times and costs along the supply chain. As long as the volume of commodity shipments did not exceed certain limits, and the size and the weight of freight cars were in line with street design and grid patterns, freight transport still worked as a "maker" rather than a "breaker" of cities, speaking in the terms of Colin Clark (Clark 1959).

This property of the logistics systems has been changing first once freight distribution followed commercial and industrial corporations in their historical move toward suburban locations. The more developed the urban fringe was becoming, the higher was the need for services in order to organize the goods and materials supply for businesses and households.

> At the turn of the century [the truck] had been a humble, little regarded conveyor of heavy loads, slowly and noisily lumbering between station and warehouse. Twentyfive years later, it had not only acquired its own unique form and status, but it had acquired the power to alter the built environment, helping to determine the location, plan, and effectiveness of many commercial buildings. (Jackson 1992, 21)

This particular contribution of freight distribution and logistics to developing, and later urbanizing, certain parts of the urban area, particularly suburbia, had been overlooked in the past, once urban studies mainly emphasized the role of the passenger car in pushing settlements beyond city limits. However, Rae who coined the new logistics hubs that replaced the warehouse a "transit shed", had simply stated: "Motor carriage not only encouraged suburbanization but also influenced the form that it took" (Rae 1971, 251). With the rising enrichment and emancipation of the periphery from the old centre, suburbia took over the function of the interface between city, region and places beyond, thus becoming a major hub in terms of logistics and freight distribution.

Suburbanization, Industrial Development and the Rise of the Poly-centric Region

Recent Patterns of Suburbanization in Germany

Consequently, any exploration of the relationship between logistics and urban development has to reflect that recent urbanization was predominantly shaped by tendencies of spatial de-concentration – which applies for the majority of the highly industrialized countries and lasted for decades. This was also the case in the Federal Republic of Germany for the period following World War II, as it was in North America, where suburbanization began some decades earlier. The de-concentration process affected the large agglomerations where out-migration of population and employment created extended suburban zones around the central cities (BBR 2005a, 191ff.). Also, an increasing de-concentration of economic activities took place, partly as reaction to population suburbanization (in the case of household oriented services), and partly caused by the intrinsic locational dynamics of certain economic activities like for instance manufacturing. Also space consuming activities like wholesale trade and logistics exhibited already in the 1970s a preference for suburban locations with good accessibility (Hesse 1999). High level producer services on the other hand remained more strongly attached to the city centres with certain exceptions like the Rhine-Main Region and Stuttgart (Eisenreich 2001).

Since the 1980s the growth dynamics in the large agglomerations have been shifting gradually from the old cores to the urban fringes and the rural surroundings (Hesse and Schmitz 1998; Schönert 2003). Medium-sized cities beyond the metropolitan areas began to form their own suburban rings. Central cities and surrounding areas merged into functional urban regions that form the spatial basis of daily activity systems for a majority of the population. This process varied in different metropolitan areas, depending on the specific historical and spatial settings: mono-centric metropolitan areas such as Hamburg or Munich showed spatial patterns different from poly-centric regions like the Ruhr, Rhine-Main, Rhine-Neckar or Stuttgart, where typical suburban locations had been traditionally mixed with older centres. The Berlin metropolitan area, where the division of Germany had formed two separate territories, presented a special case: for different political reasons, suburbanization processes predominantly took place within the city boundaries until reunification in 1989/90, particularly in the western part of Berlin.

The process of unification in Germany in 1990 represented a step forward to suburbanization dynamics in Germany (Siedentop et al. 2003; IÖR et al. 2005). It led to an accelerated suburbanization especially in East Germany that had persisted until the end of the 1990s. A major rationale for this acceleration was a lack of regional planning guidance to limit the land offers of suburban communities and also fiscal incentives for new housing construction as well as restrictions on inner city construction, due to unsettled claims for property restitution. These factors steered a large portion of the demand for housing and retail facilities to the outskirts. Since the end of the 1990s, suburbanization dynamics have been declining significantly and came to an almost complete stop in East Germany, except in the Berlin metropolitan area. Some East German urban regions even reported a reversal of the migration direction in favour of the central cities (Herfert 2002, 338). This reversal is likely to be more than just a brief cyclical interruption of a continuous de-concentration tendency. In West Germany the de-concentration process continues, but its focus has shifted from the outer suburban areas to the urban fringes, that is to say, closer to the central city (Siedentop et al. 2003). Counterurbanization tendencies that were still noticeable in the 1990s had stopped, and the overall intensity of suburbanization diminished. Since 2000, the large West German cities have again a positive population development.

With the expansion of the settlement and commuting areas, the system of the settlement structures and the central place hierarchy had changed as well. The growth of the commuting areas often followed the ideal-typical curve of land prices

(Motzkus 2002). A more or less economically rational behaviour of actors, who were attracted by low prices for rents and real estate, is generally regarded as a central impetus for suburbanization. Regarding the supply side, growth strategies of the suburban communities with extensive supplies of land for development that made regional planning controls inefficient have to be mentioned (Aring 1999). While accessibility was an important factor for suburbanization, the negative effects of high traffic volumes are regarded as one of the most pressing problems of suburban areas today. It was also criticized that the once sharp phenomenological distinction between the spatial categories of "town" and "country" is increasingly blurred. The adjustment of living conditions and concomitantly of the spatial settlement structures is, however, an almost inevitable consequence of modernization: the more suburbia appears "mature", i.e. the higher may the settlement densities of suburban locations become, the more heterogeneous their social structures are. Therefore the supplementation of residential uses by other functions is more likely. In this context suburban areas begin to resemble original properties of the city.

Summing up the tendencies outlined above it can be stated that suburban areas experienced a substantial – if regionally differentiated – revaluation over the last decades. They did not separate functionally from the central cities but have become integral parts of newly formed larger urban regions. The different parts of such urban regions are increasingly differentiated and selectively used in the course of what might be called a "regionalization of daily life": Housing takes place in the countryside or in the city, depending on income and certain phases in the life cycle. Labour is situated either in suburbia or in the inner city and spending one's leisure time is going on both in suburban areas or in the metropolitan cultural centres (Priebs 2004). Thus the spatial fix-point of the organization of everyday life is no longer the city centre, but are the individually shaped networks of activities which may stretch over the entire urban region and beyond. Urban research and regional planning reacted upon these changes by designating new concepts and new spatial categories. The "Raumordnungsbericht 2005" (Federal Report on Spatial Planning and Development) introduced the new spatial category "Zwischenraum" (intermediate space), which is positioned between the "Zentralraum" (central space) and the "Peripherraum" (peripheral space), which is characterized by specific properties concerning centrality, population potential and accessibility (BBR 2005a; BBR 2005b). Spatial categories that cover the suburban areas are "outer central space" and "intermediate space with agglomeration tendencies". Using these spatial categories, 33.9 per cent of the population and 31.4 per cent of the total employment can be considered suburban in 2003 (ibid.). A different approach by Siedentop et al. (2003) defined a radius of 60 kilometres around the centre of an agglomeration as being suburban. Subtracting the central cities from the total area inside this circle, it was estimated that about two thirds of the population lived in the suburbs and about half of the employment was located there (ibid.).

Sub- and Ex-urbanization in North America

The United States and Canada are often considered the prototypical cases of suburban development. Without neglecting the roots of modern suburban housing in the Victorian England (Fishman 1987), the model of spacious living in a single-family home became famous primarily after being adopted by a large portion of the middle-class in the U.S. (Teaford 1986). In comparative perspective, and given that there are basic commonalities as well, suburbanization in Europe and North America is characterized by divergence in terms of the extent and size of suburban dwellings, density and mixed use of urban design, and average distances e.g. in terms of daily travel. Being important determinants, the planning system and the socio-economic framework and cultural attitudes appear quite different.

The core process of suburbanization in North America was constituted by three different, interrelated developments: 1) the dispersed urbanization of the periphery in a certain distance to the core city; 2) the decline of the old downtowns, from which many of the now suburbanites had flown; 3) the emergence of new centres in the periphery often labelled as "edge cities". According to the classical studies by Douglass (1927), Binford (1985), Jackson (1985) or Fishman (1987), North American suburbanization already took off in the mid nineteenth century. Even if significant parts of suburbanization started as the selective movement of households from the core cities into the suburbs, the shift of retail and industry was much more than consecutive. As the study of the history of urbanization has put forward recently, there has always been an independent and immediate occupation of suburban space as well. The two processes of outward movement on one hand and direct suburban locational choice on the other hand were periodically connected. This was particularly true for the development of classical suburban settlements and industrial suburbs (Taylor 1915).

More recently, jobs that have either been moved to, or newly created in, suburbia, appear more advanced than before and are no longer confined to space extensive-routines in "back offices" or even mere warehouses. Retail and office functions and modern high-tech manufacturing have also occupied suburban space. As already mentioned, new centres emerged at the intersections of major highways on the edge of urbanized areas without almost any connection to the core city, yet, with a tendency towards a more self-reliant urbanization. Such new urban nodes have extensively decoupled from the old core cities, looking like something completely new rather than being just an extension of a suburb. They are found almost around all major metropolises of the U.S.; Garreau (1991) once identified 17 in the greater New York area, 16 in the Washington D.C. area. The Santa Clara County in the south of San Francisco or Orange County in Southern California represent poly-nuclear forms of "edge cities". Meanwhile almost 90 per cent of all U.S.-retail sales are being made in suburban areas and out-of-town- or edge-city-centres. These centres host two thirds of the entire North American office occupation, with about 80 per cent being provided in the last 20 years. The emergence of edge-cities has transformed the image of suburban areas significantly, even if the thesis behind was quite controversially discussed.

Edge-cities and suburban agglomerations are mingling together building a poly-centric urban landscape, perhaps among the most dynamic parts of recent North American urban development. This "post-industrial Metropolis" consists of housing areas, schools and hospitals, malls and shopping-strips, office parks, industrial parks and theme parks, in a more or less loose association. The spatial scale of these semi-urban regions is no longer the "… block-wise street pattern, but the growth corridor stretching along 50 or 100 miles" (Sudjic 1993). Lang (2003) has recently coined this type of settlement the „edgeless city". The target point of immigration into these places is no longer the core city, but the entire topographic space.

According to the latest U.S. Census, the 90s-decade saw a population increase by 13.2 per cent up to 281.4 million people and was the only one during the twentieth century that witnessed population gains in all U.S. States (U.S. Census Bureau 2001; Katz and Lang 2003). Whereas big cities such as New York or Chicago experienced population gains as well (some of them for the first time in decades), urban regions in general were growing in multiple directions: many suburbs were urbanizing, whereas their edges kept further expanding. The reasons behind this strong growth are to be economic growth and immigration. Population concentrations have been shifting in the second half of the last Century from the Mid-West and North East toward regions in the southern and western parts of the country. Thus core cities were stabilizing in terms of population and occupation or even performing better than before, and the old suburbs became mature, due to specific life-cycle effects such as demographic change, immigration and social diversification. In this context, Frey (2003) had already observed the emergence of certain "melting pot suburbs". This particular pattern of urbanization of the suburbs is both accompanied by positive and critical impacts, e.g. in terms of diversity and density, but also regarding infrastructure bottlenecks, loss of open space and punctual decline. Besides a constant demand for places that accommodate a single-family house oriented lifestyle (Rich 2001), a remarkable shift of population is also directed towards exurban and rural areas. These are predominantly located at greater distances from agglomerations. A prototypical case in this respect is the State of Nevada that absorbs much of the growth that emanates out of the metro-regions in Southern California and the San Francisco Bay Area. Compared to the economically scarce and politically highly contested land use policies in the coastal regions of California, the inland offers cheap land, infrastructure and – increasingly – those amenities that the apparently urban communities demand for. To a significant extent, Kotkin's "New Geography" (2000) consists of highly qualified exurbanites, employed in the IT and New Economy sectors who do not necessarily need to be present at their workplace all time. The more remote "leapfrogging" is made possible by the provision of information and communication networks the more they can afford to commute, but only periodically. In contrast, middle-class people are increasingly pushed out of the metro-regions due to high rents and land prices, but actually can't afford the daily commute to their offices.

Based on the assumption of constant socio-economic framework conditions, the two major trends of the urbanization of the suburbs and the development of more remote, exurban areas may remain stable in the foreseeable future. One of the resulting types of settlements is the highway-corridor that stretches between initial nodes of urbanization that later on tends to grow together. This phenomenon can be observed in many of the newly emerging growth regions, e.g. in the Southeast or Southwest of the U.S.:

> Bands of Suburbs have started to merge with each other along Southern Transportation Corridors, in some cases forming almost unbroken chains of medium-density areas hundreds of miles long – from Atlanta to Raleigh along Interstate 85, or from Washington to Norfolk." (*The New York Times*, 17 April 2001)

The interpretation of such patterns is still controversially discussed: Whereas historians like Robert Fishman now believe to observe a kind of urban renaissance, following the past age of suburbanization, others expect the coming of a suburban megalopolis different in form and size that never existed before. However, there seems to be a consensus that traditional terms of urban research may no longer be appropriate for conceptualizing these developments: "The geographical definition of suburb is outdated and has to be changed" (*The New York Times*, 17 April 2001). This assumption is for a long time supported by empirical evidence on the emergence of suburban employment centres – not confined to typical "edge cities", yet representing a more general pattern of development. Since the 1980s (Cervero 1989) a stable de-concentration of occupation has contributed to a long term convergence of supply of and demand for jobs at different places outside the metro-regions. This effect is associated with a stabilization or decrease in average commuting times (Gordon et al. 1988; Gordon and Richardson 1997). This event seems to confirm earlier ideas of an "urbanization of suburbs" (Masotti and Hadden 1973) and a "changing face of suburbs" (Schwartz 1976), or simply a "new suburbanization" (Stanback 1991). It is characterized by offering the same advantages of agglomeration as the core cities did before. In the 1980s Muller has coined this successive development of the suburbs as a "transformation of bedroom suburbia into the outer city" (Muller 1989, 39). In the context of an evolutionary perspective on urban development, this discussion has also contributed to a changing assessment of suburbanization.

Commercial and Industrial Suburbanization

Commerce and industry as drivers of spatial de-concentration have been overlooked in the scientific investigation of suburbia to a significant extent. For a long time, the core icons for explaining suburbanization particularly in North America have been attractive housing (ownership, garden-like), the search for "community" and a safe environment, and the pro-active policies of planning and infrastructure provision, e.g. by loans and cheap land, accessibility, water and waste management (Hayden 2003). However, the constant change of locational dynamics in industry, retail and services can be considered a continuum in the urbanization process both in Europe and in North America (Hayter 1997). This applies for the emergence of a differentiated spatial division of labour and specific urban structures in suburbia. More recent research has revealed that the suburban commercial landscapes had been established earlier than assumed before, that they did not solely follow the decentralization of housing, and that they were changing according to a specific cycle of development (Lewis 1999; Walker and Lewis 2001). As a consequence, the traditional image of commerce and industry as prime agents of suburbanization have been changed as the social science did before, related to the recognition of suburbia as a complex place of individual and societal practices. Considering that, conventional images of suburbs as bedroom communities and appendices of core cities have to be rejected:

> In the classic studies, suburbia is conjured up as an image of "homes in a park"; a middle landscape constituted as a way of life halfway between city and country. This conventional wisdom needs considerable revision. Residential areas have not singularly led the way outward from a previously concentrated city, but have always been joined at the hip by industry locating at the urban fringe. The outward spread of factories and manufacturing districts have been a decisive feature of North American urbanization since the middle of the nineteenth century. (Walker and Lewis 2001, 3)

In the modern industrial city, the location of industries was placed more or less close to the centre of the city: not exactly within the city centre (that was primarily devoted to the most productive or most symbolic land uses), yet, adjacent to the centre. These areas were easily accessible by workers or offered direct access to a transport infrastructure (port, railway). As mentioned above, in the classical concentric model by Burgess, this area was dedicated "zone of transition", in the two other theoretical models of social ecology (Harris and Ullman 1945; Hoyt 1939), these were the separately localized light industries and heavy industries. Both the sector model and the model of the poly-nuclear city have already included or anticipated the shift of industrial land use out of the core city: according to rising industrial growth rates, particularly with further mechanization and automation of production supported by the progress of mobility technologies, a new freedom and a new imperative of locational choice could be found. Firms had been using this opportunity for developing and industrializing space. This process recommenced in Europe and in North America during the mid-nineteenth century, initially at a limited pace but continued more quickly afterwards (Kostof 1993). It is prototypically visible in the historical districts of the City of Berlin, where the machine tools-industry (e.g. the manufacturing of railway locomotives) was moved stepwise from the centre via the north of the city to Tegel a district in the northeast.

These processes of establishing industrial suburbs were extremely important for the growth of the entire region, yet they have been significantly overlooked by urban researchers, particularly due to the lack of data, and also because of the frequently practised annexations that concealed suburban developments, now being analysed as if they occurred within city limits (cf. Lewis 1999, 149f.). By comparing the developments in central England, the Ruhr Area in Germany or in the industrial belt in the American Northeast and Midwest, it looks as if industrial suburbanization continued in a regionally differentiated manner. However, there are commonalities that can be related to three issues: the establishment of certain types of suburban industrial districts, the driving forces for the shift of commercial land uses out of the core cities, also the constant change of suburban locations under the influence of service industries and new technologies.

In more generic terms, the suburbanization of industry has brought up different functional types of land use, depending on the major players within such processes and where they did happen (Lewis 1999, 161).

The first type considers the incremental, informal expansion of existing locational structures, which resulted predominantly on the edge of the core city. It concluded with the infill of existing industrial areas and was also a consequence of planned shifts of the firms that aimed to avoid locational disadvantages in the centre (e.g. the degree of workers" attachment to labour unions). Also new resources in the wider region were supposed to be used. This form of urban expansion by industrial location choice is considered by Lewis (ibid.) being the most popular and most frequent in the past. The same might basically apply for the suburbanization of commerce and industry in Europe.

In contrast to the first type, Lewis makes explicit reference to Taylor's view on the industrial "satellite city" (Taylor 1915), which included a separate unit in some distance from the core city, which was e.g. the case with the "Siemensstadt" in Berlin. Although these satellites had initially been planned and designed as autonomous entities, they were depending on the core city and have afterwards coalesced with other subdivisions and thus constituted the metropolitan region. The satellite city was relatively diverse and multifunctional, developed an own labour market, a retail catchment area and also proper traffic connections.

The third type had been developed as a "company town"-suburb. The driving force that created such places was a single corporation in most cases; prototypically, Gary or Pullman in the Chicago, Illinois, area are known in this respect. The related, separate firm locations were extremely huge and normally not integrated into the existing built environment. A common background of many company towns was experimenting and applying new methods of production organization, which often required flat buildings on a single lot; thus the decline of the multi-story manufacturing building was associated with changing locational requirements.

The fourth type is represented by the organized industrial district. It was accessibly located at the urban fringe. It was mainly established due to cheap land and the offer of "amenities": certain advantages in terms of tax rebates, a special design of the district, financing and other supportive measures. The districts were particularly targeted by SMEs that were seeking for a "safe haven" for their manufacturing plant (Lewis 1999, 161). In many cases, the district was built up by real estate developers or railway companies; in this respect, it is considered a predecessor of the commercial or industrial parks of today.

The dynamic process that helped develop a more urban suburbia and also a differentiated industrial landscape has been coined as "geographical industrialization" or "industrial dispersion" (Storper and Walker 1989, 99; Walker and Lewis 2001, 6); this theoretical frame will also be used in the following as a reference point for studying and interpreting locational dynamics of logistic in suburban areas. This approach appears to be useful because of its dynamic view of the interrelated factor combinations and corporate strategies that establish suburbia in the context of the poly-centric urban region. Such framework of factor conditions that determine the movement of firms towards suburbia can be considered both the result of changing general framework conditions and regionally specific responses. They consist of a set of factors, first of capital driven, expansive growth of firms, moving beyond the limits of existing locations; second innovations in production organization that demand for larger, more appropriate sites; third problems of accessibility at the traditional interfaces of the core city (ports, freight railyards); finally changes in the regulation of industrial relations, particularly the evolvement of power of labour unions, that made core city firms move to peripheric places which were not yet unionized. Their locational choice also took place in suburbia due to an institutional frame that ensured the provision of cheap land and amenities. In this regard, both in Europe and in the U.S. the most active agents of suburban development were municipalities and "terrain"-developers.

After World War II, industrial-commercial suburbanization in Europe became further developed, reinforced and differentiated in a dynamic way. Meanwhile, the majority of employment in Germany had been settled at suburban locations. This applied for manufacturing which performed particularly better, as firms had been moving from the core agglomerations to both urbanized and rural locales. Also, empirical evidence reveals that there is a qualitative change underway, carried by the emergence of services instead of manufacturing. This is true for corporate services, particularly manufacturing-, processing- and logistics-oriented services, such as wholesale and distribution, and also even more premium services that had been confined to the core city until then:

> Commerce and industry are going to become pushed out of the city. This leads to a shift to the near suburbs (first ring), which are itself targeted by the service industries. competition is even more strongly here, due to the presence of several groups of agents, like established service firms, manufacturing firms, newcomers and suburban movers. Therefore as a consequence, commercial developments are more and more shifted towards the second or third ring within or beyond the agglomerations. (Kahnert 1998, 510)

This cycle-wise creation of localities by specific users, their competitive pressure and the transformation and re-development of locations constitutes the more recent commercial landscapes of suburbia; thus suburban locations are subject of a comparable competitive pressure as the industrial areas in the core cities had been exposed to earlier. This is an outcome of a changing economic framework (globalization, large-scale network embeddedness of firms) and changing modes of technology and organization (new modes of production, supply-chain oriented and modular processing). Finally, it is a result of the specific ability of suburban locations compared with the core city. A respective structural change has happened in all industrialized, capitalist economies since the 1980s at latest. In post-socialist transformation economies such as Eastern Germany it became even more accelerated, where light industries and distributive land uses primarily searched for places close to the motorway network (Usbeck 2000).

The Poly-centric Region

The process of suburbanization in Germany has contributed to the mere expansion of agglomerations and to the emergence of poly-centric urban regions, which in German is also conceptualized as the "Stadtregion" (cf. Boustedt 1975). To some extent, these urban regions have replaced the traditional dichotomy of core city and suburbs in Germany as well. Their spatial structure is characterized by tendencies of sub-centralization and by further dispersion. The related changes have been analysed in general, e.g. In the case of the urban regions of Hamburg, Frankfurt, Stuttgart and Munich. In such regions, economic growth, social differentiation and spatial structure have been connected in a special way (cf. BMBau 1996). As discussed earlier, related U.S. developments appear much more representative. In this respect, the two case study-regions Berlin-Brandenburg and the Bay Area/the Central Valley share the property that they are no longer dominated by a core city, but deployed dispersed, highly differentiated patterns of land use and interaction. The poly-centric urban region is defined in contrast to the old, mono-centric model of the city which belongs to the following properties: the existence of several centres, the no longer strict distinction between city and suburbs, the increase of services compared to manufacturing, information technologies, and the key role of mobility, transport and logistics (Kloosterman and Musterd 2001, 623). The genesis of the poly-centric urban region did not only occur because it was based on the growth from the centre to the periphery, but also as a consequence of endogenous suburban growth. In this respect, the suburbs contribute to the emergence of the poly-centric region:

> Suburban growth as a whole has been a mixture of industry and homes, the city sprawling ever outward from its initial point of establishment and repeatedly spilling over political, social and perceived boundaries. The result has been extensive, multinodal metropolitan regions. (Walker and Lewis 2001, 3)

The poly-centric region is usually larger than the old city had been; it is much more complex and diverse. It is much more difficult to manage in terms of policy and planning, which might deconstruct any conventional wisdom that aims at constructing the "region" as a political product and concept. The decisive difference I suggest distinguishes the poly-centric region from the old core city including its suburbs is the matter of fact that the urban region does not only deploy a differentiated *inwards* pattern of relationships and interaction, yet is also connected outwards in a new way. Mobility, transport and logistics are thus playing a major role in the genesis, enrichment and stabilization of the urban region. Transport accessibility and household motorization do not only allow to spatially expand individual lifestyles and insofar a "regionalization" of lifeworlds. They are also becoming the driving force of a new mix of long-distance and local mobility. This also fundamentally questions the classical categories of distance-based or -dependent behaviour in research and planning.

If these processes are analysed in the context of a long-term assessment, a cycle of expansion and contraction of the urban space seems to be prevalent. This cycle puts the traditional idea of the constant growth of cities into perspective. In this respect, the phase-oriented model of urban development (van den Berg et al. 1982) seems to be quite reasonable, as long as it supports the view of ups and down rather than assuming a certain finalized development. The latter argument was also the subject of substantial criticism, since the model was assumed to follow a deterministic perspective of urban development that neither fits with poly-centricity nor de-concentration in a satisfactory way. However, empirical evidence supports the argument that German agglomerations have mainly emerged from suburbanization and city-region building, yet that there are no vital signs both for de-urbanization and dissolution of the city or, in the other extreme, re-urbanization and renaissance of the core city. Given the situation in the U.S., it seems more appropriate to conceptualize a mosaic of sub- and ex-urbanization.

As long as urban regions as a whole are winners, not losers of contemporary socio-economic and spatial change, the case is about the poly-centric ("city"-) region as a result of modified urbanization, not the decline of the city – what else would explain the constant forces of attraction underlying agglomeration? This modified pattern of urbanization however includes remarkable qualitative differences compared to the classical understanding of urban development or suburbanization, according to Table 1.2.

At the large-scale level, the relational logic of the poly-centric region appears to be equivalent to network developments in economics in general and in transport and logistics in particular. As the traditional, resources and materials based paths of industrialization in the nineteenth century were highly important for urbanization, contemporary structural changes contributed to a de-concentrated pattern of urban and regional development. These forces of de-centralization seem to be much stronger than they were during the twentieth century's suburbanization. The material fragmentation and symbolic dissolution of the city into the region is comparable to the de-composition of the factory in favour of flexible production

Table 1.2   Urbanization and spatial development

	Urbanization	Suburbanization	Post-suburbanization
Settlement structure	Absolute and relative increase of urban population and occupation	Decline of the core city, rise of suburbs, emergence of the "Stadtregion"	Qualitative change of suburbia, increasing significance of the periphery (i.e. in rural areas)
Interactions in space and time	High internal orientation, city is the place for housing, labour, leisure	Intensive exchange between city and suburb (e.g. by commuting)	De-coupling of land use and spatial interaction from the old core city
Socio-cultural dimensions	Urban lifestyles, urbanity, the city as the locale of politics	Urban lifestyle in a green setting, NIMBY-syndrome	"Hybrid" lifestyles in urbanized villages or ruralized cities
Symbolic means, myths, social construction	European City, public space, compact city	"Urban sprawl", bedroom suburb	"Zwischenstadt", [edge city, exurb]

Source: own

units and value chains. It is not coincidental that the network metaphor has gained increasing popularity both in economic sciences and in geography and urban studies. Hence a major research question of this study will be related to the potential link between network structures of settlements and those that are emerging in logistics and supply chain management. Such considerations translate into the question whether the traditional function of goods supply, once responsible for the emergence of the city, is now contributing to the decline or at least the transformation of the city.

Suburbia Organization Space

According to recent observations, the traditionally strong ties between the city and economic development in general and freight distribution in particular develop in a differentiated way. On one hand, the historical link between urbanization and accessibility seems to be stronger than ever: the functionality of logistics – in a broad understanding of the term – appears as a prime location factor for cities in the changing framework of the network economy or network society. "Cities and urban regions become, in a sense, staging posts in the perpetual flux of infrastructure mediated flow, movement and exchange." (Graham 2000, 114) The new role of services or the upcoming demand for "agile" manufacturing might not be accomplished without the appropriate logistics and infrastructure basis, as does the increasing inter-regional trade. On the other hand, both in a metaphorical and literal sense, there is an emerging notion of "flow" in contemporary urban and geographical thinking. According to relational approaches in economic and social geography, the development of place is increasingly embedded in *relation* rather than *fixity*, in *routes* rather than *roots*. As it is expressed in terms like "Terminal Architecture" (Pawley 1994) or "Flow City" (Graham 2001), this language reveals a certain notion of volatility and dissolution. Accordingly, the ability of urban places to act as a spatial fix in an environment of increasing flows can indeed be questioned.

One of the leading conceptual approaches in this respect had been developed by Manuel Castells. According to his considerations on the "Network Society" (Castells 1985, 1996), Castells suggested that the informational "space of flows" might somehow determine the traditional, physical "space of places", driven by innovations in communication and information technology:

New technologies allow the emergence of a *space of flows* substituting for a *space of places*, whose meaning is largely determined by their position in a network of exchanges. The logic of large-scale organizations fits perfectly into a spatial form that abstracts from historic reality and cultural specificity to accommodate new information and instructions. (Castells 1985, 33)

Thus the term of the "space of flows" can be used as a framework for relating the new information and communications technologies to the spatial dimension. The respective system of inter-relations refers to the idea of the global network, consisting of nodes and linkages, much better than compared to the traditional understanding of the city as a market place.

In this respect, contemporary interpretations of urban places as economic spaces have gone far beyond the meaning of the city as a mere concentration of people, jobs or added value. Cities are increasingly considered being a part of a large-scale economic network, rather than a spatial fix in the world of flows:

Instead of conceiving cities as either bounded or punctured economic entities, we see them as assemblages of more or less distanciated economic relations with different intensities at different locations. Economic activity is now irremediably distributed. Even when economic activity seems to be spatially clustered, a close examination will reveal that the clusters rely on a multiplicity of sites, institutions and connections, which do not just stretch beyond these clusters, but actually constitute them. (Amin and Thrift 2002, 52)

Consequently, the authors "… replace the idea of the city as a territorial economic engine with an understanding of cities as sites in spatially stretched economic relations" (Amin and Thrift 2002, 63).

However, the widely dispersed and disconnected parts of this system need to be re-connected. Adding to the layers of the immaterial flows of knowledge, information and finance that are bound together in urban economies, these material connections are taken over by transport and mobility, by logistics and freight distribution. However, logistics organization needs "organization space" (Easterling 1999). Easterling assumes that contemporary urban developments such as transport interchanges, ports, airports, malls, economic franchises can best be understood as dynamic sites for organizing logistical processes. "The primary means of making space consist of a special series of games for distributing spatial commodities". (Easterling 1999, 113) She also points out that the critical architectures of these spaces are not visible, but are woven into their extended technical and information systems and often hidden infrastructure networks.

> The real power of many urban organizations lies within their relationships between distributed sites that are disconnected materially, but which remotely affect each other – sites which are involved, not with fusion or holism, but with adjustment. (Easterling 1999, 113)

Such sites that tend to resemble urban functions in terms of connecting economic processes and places are exemplified by Amin and Thrift (2002) in the case of the five global distribution centres operated by Eastman Kodak. They are considered being key nodes within a newly emerging, flow based pattern of creation of value:

> Distribution centres [...] are carefully scattered around key urban gateways and transport nodes, marking a geography of delivery that has virtually no connection with the original geography of production and the final geography of consumption [...]. (Amin and Thrift 2002, 69)

It is still controversially discussed whether these de- and re-connected infrastructure hubs do represent a broader trend in urban and economic development. However, complex interdependencies of the local and the global as a result of large-scale network architecture and embeddedness seem to be the norm rather than the exception in contemporary economic development. Even if empirical evidence does not support the assumption of a general "footlooseness" of the economy, the freight sector reveals an astonishing degree of disconnection of logistics networks from traditional urban and economic network topologies. Would the city, once prime market-place and site of economic exchange, become transformed to a mere terminal, providing the transshipment of commodities from A to B, without earning a certain added-value that traditionally emanated from freight handling?

Chapter 2

Technocapitalism and Logistics Transformation

A Changing Economic Environment: New Technologies and Globalization

Logistics includes "the process of planning, implementing and controlling the efficient cost effective flow and storage of raw-materials, in-process inventory, finished goods and related information from point of origin to point of consumption for the purpose of conforming to customer requirements." (Council of Logistics Management 1986) Although logistics initially emerged from the preparation and organization of military operations, it is now used to optimize the functions of production, distribution and consumption in businesses and corporations. The offspring of modern logistics and freight distribution sector goes back to the modernization of the capitalist economy, the development of specific modes of industrial production and the emergence of a particular division of labor. This created a distinct "sphere of circulation", situated between production and consumption (Marx 1939/1953). To a certain extent, it was the system of material circulation that allowed for the transition from use-value to exchange-value, which made the large-scale capitalization of commodities possible. Mass distribution and marketing later became incorporated in the practice of modern management (Chandler 1977) and it also became increasingly specialized and less dependent from production or consumption.

The recent development of logistics and physical distribution is an outcome of a broad set of economic structural changes and the related corporate responses. Among these changes, several components have gained particular importance, regarding the questions what commodities are going to be produced (distributed and consumed): how this is being operated and where it happens. First, sectoral shifts including the rise of service economies, the increasing share of goods with high value and low weight and also the upcoming high tech and knowledge based sectors play a major role in shaping the economic environment (IMF 2001). These interrelated changes supported high tech and knowledge based sectors (Castells 1996). The rising service economy in general is also one major factor that helped establishing logistics and freight distribution services in particular. Logistics, in turn, enabled regionally bound economies to further dissociate in terms of space, time and organization.

Second, the introduction of new information and communication technologies allowed the integrated management and control of information, finance and goods flows and made many of the aforementioned changes possible. A major

requirement of the implementation of physical distribution was the invention
of modern information and communication technologies (ICT). Features like
electronic data interchange, the automatization of product flow in dedicated
warehouses and distribution centres (DCs), computerbased tracking-and-tracing
systems (which offer on-line insight into the status of your shipment via the web)
or the recent, radio-frequency based identification (RFID) are primary sources
of enormous productivity gains over the last two decades. The same did the
standardized container much earlier which industrialized trade and transport from
its offspring in 1956. An important requirement for the successful implementation
of these innovations was the expansion of hard infrastructures such as highways,
terminals and airports. Consequently, the material foundation for the today so
called network economy is based on both: modern logistics and on asphalt and
concrete.

Corporate strategies in physical distribution were primarily associated with
two technological innovations which were developed in the late 1970s/early 1980s
and almost revolutionized logistics: the bar-code and electronic data-interchange
(EDI). Both allowed a completely new operation of inventory management,
adjusting the entire material management according to customer's demand and
thus reducing inventory costs substantially. The main disadvantage of bar-code-
based inventory management and particularly EDI-appliances was that they are
very expensive and not standardized. Both reasons led to a limited degree of
implementation, because only larger firms could afford to invest in these systems.
With the advent of the Internet, the situation changed dramatically: prices fell
down and acceptance increased accordingly, whereas the "http-protocol"
functioned as the main common standard for all web-suppliers and users. Now
the web has become a major rationalization tool, not only for individual firms and
their subsidiaries, but for managing the entire producer–customer relationship.
The accelerated pace of technological innovation contributed to further changes
for instance by introducing global positioning systems that allow for identification
and routing of vehicles in a way that was not possible before.

Third, globalization has changed the framework of development, growth and
economic exchange significantly with an emerging network of global flows and
hubs that depends upon efficiently working transport systems and infrastructure.
Despite the degree of uncertainty regarding the precise origins, nature and
impacts of globalization, there is growing consensus that global integration is a
permanent and long-term trend that has contributed to fundamental changes and
triggered a variety of responses. Insofar it is widely accepted that globalization
can be best understood as a fundamental transformation consisting of a variety
of processes and impacts (cf. Held et al. 1999, 3). According to Holton (2005),
three basic characteristics make up the shape of what we currently understand as
globalization: 1) the rising exchange of people, goods, information, values, habits,
as it is particularly indicated by global air travel and maritime shipping by trade
statistics or by an increasing level of social and political interactions, but also
by numerous contentions; 2) the increasing degree of interdependence of nation

states or national economies, which is particularly evident if looking at the supply, production and distribution of manufactured goods across the globe; 3) the rising awareness in societies that the world is a closed system which has to be shared by the global population, that natural resources appear to be limited (such as oil) and that environmental risks can have wide-scale impacts (such as the issue of global warming). It is also evident that globalization is not neural in terms of space, thus being associated with the dissolution of material space into the virtual world of information transfer, but that there is a structural demand both for material connections between physical places and for the infrastructure that is required to enable such connectivity.

The structural re-arrangement of value chains and production networks on a much higher spatial scale than before has evolved in the emergence of global production networks (GPNs). GPNs developed as a consequence of innovations in information and communications technology and of the increasing degree of global economic and social integration associated with globalization in economic, social and cultural terms. GPNs have been established to cover major parts of the globe at a very dynamic rate in countries recently integrated into the new geography of global production. As Dicken (2003) has put it in his seminal "Global Shift", the establishment of GPNs no longer occurs in traditional, natural-resource-based and labour-intensive branches such as the apparel industry, but also in highly competitive, modern industries such as electronics and computers (including components), machinery or automotives, as well as in consumer-related services such as retail and wholesale. Thus GPNs reinforce the general trend of increasing global trade rates.

Globalization also indicates a major shift in the operation and location of manufacturing compared to earlier periods of global integration. Initially, a new international division of labour assigned assembly lines and their operation to developing countries, because of their cheap labor and much lower degree of regulation compared to the so-called developed world. Value-creating activities such as research and innovation, product design and marketing had been reserved for the home countries and home markets of global corporations. The more recent character of GPNs suggests that subcontracting and production-related services are being fragmented and shifted towards – what used to be called – periphery. This is also true for core service industries such as data processing, software development and call centers, and for retail operations. As Coe and Hess (2005) have noted with regard to retail, these businesses are thus increasingly becoming transnationalized, and economic globalization is reaching a degree of global integration not known before. Driven by global integration, GPNs are also likely to foster regional development since every global network tends to be embedded in local and regional places. Coe et al. (2004) have emphasized this point with respect to the often acclaimed interaction between the "global" and the "regional": it is rather a close interrelationship than a dichotomy between global and regional processes emerging out of GPNs. However, it is obvious that logistics and physical distribution are major requirements for organizing such globalization of trade and manufacturing and the emergence of GPNs.

The changing framework conditions as they were addressed above including sectoral shifts, new technologies and globalization, have triggered fundamental changes in the organization of production, notably manufacturing, that had been introduced to most of the highly industrialized countries since the late 1970s and early 1980s. Logistics plays are somehow heterogeneous role in this respect, since it was first driven by the changes in the production environment and committed to organize the respective, derived material flows. Later on, it became more or less integrated and "structural" developing into strategic means of spatio-temporally adjusting the entire process of creating added value (see below).

There are several paradigms under which these changes in corporate and interfirm organization, particularly manufacturing, have been discussed. Speaking in more generic terms, the changes emerged against the background of general organizational change in corporations (see Nelson and Winter 1982) that triggered new modes of production organization. Some of these developments were interpreted alongside the umbrella-paradigm of "flexible specialization" that was developed in contrast to the classical Fordist pattern of standardized mass production and consumption (Dicken and Thrift 1992; Gertler 1992; Piore and Sabel 1984). In more standardized manufacturing environments, "lean management" emerged as a primary mode of making production process more efficient and less costly (Womack et al. 1990; Harrison 1997). Lean management was initially invented by Japanese corporations e.g. in the automotive business during the 1980s in order to concentrate the entire car assembly process to the essential minimum of activities and resources. One of its main premises is to eliminate inventories and organizing materials supply strictly on demand, replacing the former storage and stock keeping of inventory.

As a part of lean management, "just-in-time production" (JIT) has put a particular emphasis on transport and logistics by decomposing the manufacturing process and thus generating a rising demand for shipping and delivery, particularly of smaller units in a higher frequency, an increasing importance of time, reliability or even speed, and also new infrastructure requirements both in terms of carrying flows and of operating nodes and interfaces.

> Originally, the use of the JIT practice meant that through the employment of expedited freight transportation, inventory was reduced to an absolute minimum. As the practice has matured, companies have been able to reduce both inventory and transportation costs through the use of advanced telecommunications, data management, refined production planning and increased "in-transit" visibility (which allows companies to know precisely the location of stock while in transit). Thus, companies have adjusted JIT practices by ordering stock further in advance but compensating by using less expensive transportation options. For example, instead of using overnight air, less expensive two- or three-day trucking service may be used. (Robins and Strauss-Wieder 2006, 7)

More recently, modernized forms of production organization and the related provision of logistics services are discussed in the context of the paradigm of "modular production". Sturgeon (2002) provides a comprehensive overview of this pattern of production organization. It is driven by contract manufacturing and a vertically disintegrated, horizontally integrated management of value-chains. Modular manufacturing comprises both small and large firms, small and large geographical scales, and it aims at creating a large number of products using few processes in order to receive maximum revenue through economies of scale (Sturgeon 2002, 477). Modular manufacturing, as lean management was before, is also associated with certain forms of geographical proximity, particularly concentration that appears to be efficient in terms of coping with complexity, limiting transaction costs and generating tacit knowledge (Frigant and Lung 2002). However, the application of such production paradigms, as different they are in their nature and contents, have one similarity: they were creating a fragmented, differentiated sphere of production that required an extremely efficient, adjusted system of circulation: logistics and freight distribution.

Supply Chain Management: Towards Time-space Fragmentation

In order to describe the transformation or deconstruction of the firm in the context of modular manufacturing, Suarez-Villa (2003) has emphasized the network logic of the modern economy and coined the term "techno-capitalism". It is used as an interpretative scheme for analysing structural changes that are primarily driven by information technologies and globalization. Network building, devolution of hierarchies and speed are its main characteristics. Logistics and distribution are becoming a key unit within this production system, since it requires all modules to be agile (flexible) and to interact within the network. Flexibility is both an organizational and a geographical requirement. Thus a major shift has occurred in how and where commodities and their components are being assembled, manufactured and distributed, which is depending on logistics and freight distribution, and, in turn, shaping this subsector. New services such as third- and fourth-party logistics providers integrate freight transport, warehousing, logistics and physical distribution, which occur at increasing spatial scales.

The key term in the analysis and interpretation of modern logistics is supply chain management: the comprehensive management of value added processes and interactions including all components and activities that are dedicated to the manufacturing, processing, marketing and finally consumption of commodities (cf. Gattorna 1990, 8). Different from earlier versions of logistics management, supply chain management is being understood as an integrated approach, in which logistics management is embedded. Initially, the two segments of materials supply and physical distribution had become integrated; afterwards, they became part of a broader logistics concept of the firm (Bowersox et al. 1968; Hall and Braithwaite 2001). "One of the most significant paradigm shifts of modern business management is that individual businesses no longer compete as solely autonomous entities, but rather as supply chains" (Lambert 2001, 99). The increasing degree of integration along the logistics chain that is provided by the concept of supply chain management was only possible through the invention and application of the new information and communication technologies. They made the complete material flow transparent in terms of process and costs, space and time, and thus allowed for its integrated management. As a consequence, productivity improvements were initially achieved within one factory and logistics was able to achieve between the different elements of the supply chain: a system of integrated factories. As a result, a "principle of flow" had been invented and consequently implemented in corporate organization that permitted the reduction of inventories in time-sensitive manufacturing activities from several days" worth to several hours.

Once logistics was an activity divided into the supplying, warehousing, production and distribution functions, most of these functions were fairly independent from the other. With the new organization and management principles, firms were following a more integrated approach, thus responding to the upcoming demand for flexibility without raising costs. As production became fragmented, activities related to its management were consolidated. Spatial fragmentation became a by-product of economies of scale in distribution. In the 1990s, with the convergence of logistics and information technologies, this principle was increasingly applied to the whole supply chain, particularly to the function of distribution. In some highly efficient facilities, the warehousing function went down as far as 15 minutes" worth of parts in inventory. It is now being introduced in service functions such as wholesale and retail where inventory in stores are kept at a minimum and resupplied on a daily basis.

The integrated organization of industrial manufacturing and distribution has already been pursued for some time. In this regard, even the 1960s were considered a key area for achieving major productivity improvements (Bowersox et al. 1968). However, only with the implementation of modern information and communication technologies did this assumption become possible. They allow for the integrated management and control of information, finance and goods flows and made possible a new range of production and distribution systems (Abernathy et al. 2000). Step by step and according to improvements in operating the information and communication technologies, the two ends of the assembly line became integrated into the logistics of the supply chain: the timely supply of raw materials and components from outside, and the effective organization of distribution and marketing (see Figure 2.1).

The flow-oriented mode of corporate management and organization now affects almost every single activity within the entire process of value creation. The core component of materials management is the supply chain, the time- and space-related arrangement of the whole goods flow between supply, manufacturing, distribution and consumption. Its major parts are the supplier, the producer, the distributor (e.g. a wholesaler, a freight forwarder, a carrier), the retailer, the end consumer: all of whom represent important players and particular interests

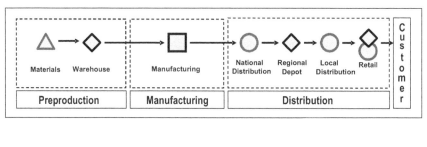

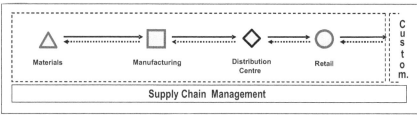

Figure 2.1   Conventional and modern (Integrated) supply chain

Source: own

(Bovet and Martha 2000; Bowersox et al. 2000). Compared with the former, more traditional shape of the freight transport system, the evolution of supply chain management and the related emergence of the logistics industry is mainly characterized by four properties:

- First, a fundamental restructuring of goods merchandising by establishing integrated supply chains with integrated freight transport demand.
- Second, whereas transport was traditionally regarded as a tool for overcoming space, logistics is critical in terms of time. This was achieved by shifts towards vertical integration, namely subcontracting and outsourcing, including the logistical function itself (Harvey 1989).
- Third, according to macro-economic structural changes, demand-side oriented activities are becoming predominant. While traditional delivery was primarily managed by the supply side, current supply chains are increasingly managed by demand.
- Fourth, logistics services are becoming complex and time-sensitive to the point that many firms are now sub-contracting parts of their supply chain management to third- or fourth-party logistics providers. These providers benefit from economies of scale and scope by offering integrated solutions to many freight distribution problems.

Summing up these tendencies, as a consequence of the growth in international trade, the establishment of global production networks, the geographical and functional integration of the subsectors of production, distribution and consumption, complex networks have been established that involve the increasing flux of information, commodities, parts, and finished goods (see Table 2.1). This in turn demands a high level of command of logistics and freight distribution. Whereas Alfred Chandler had put some emphasis on "the revolution in distribution" almost thirty years ago (Chandler 1977, 238), things have changed further. According to recent interpretations of industrial change, the "visible hand" once coined by Chandler appears to have been replaced by the "vanishing hand" (Langlois 2003) – as an outcome of more structural, system-wide activities and effects.

However, in the new competitive environment there are also powerful single actors that have emerged which are not directly involved in the function of production and retailing, but mainly taking the responsibility of managing the web of flows. The global economic system is thus one characterized by a growing level of integrated services, finance, retail, manufacturing, and nonetheless distribution, also at the level of the single corporation. A good example of these powerful actors is given by the American retailer Wal-Mart, the largest corporation in the world, which has become the "template" for late-modern capitalism (Lichtenstein

Table 2.1   Phases in the development of logistics

Phases	Subject	Aims
1970s	Traditional logistics	Optimization of separate functions
	Supply – storage – manufacturing – storage – distribution	
1980s	Logistics as cross-section function	Optimization of processes
	Supply – logistics – manufacturing – logistics – distribution – customer	
1990s	Phase of functional integration	Establishment and optimization of process chains
	Logistics integrates functions into process chains: Customer – development – supply – order management – distribution – re-cycling – customer	
	Phase of comprehensive, inter-firm integration	Establishment and optimization of value added chains
	Logistics integrates firms into value added chains: Customer – distributor – logistics provider – manufacturer – retail – logistics provider – customer	
2000s	Phase of worldwide integration of value added chains	Establishment and optimization of global networks
	Logistics integrates value added chains into global networks	

Source: own after Baumgarten/Thoms 2002, 2

2006). By linking newly emerging, competitive places that provide cheap goods production and supply (e.g. in China) with major customer markets in North America, among others, the company's profits are growing in an unprecedented manner. The case of Wal-Mart provides some evidence that, besides the efficient management of financial resources, the core reason for its rapid success as a truly global company results from its ability to manage its supply chains and networks extremely efficiently (see Bonachic 2006).

ICT and E-commerce: The Annihilation of Physical Distribution?

The Influence of ICT-use on Logistics and Supply Chain Management

Major impacts on the way supply chains and the associated goods distribution are being organized have been expected by the application of information and communication technologies (ICT), particularly in the case of electronic commerce (or E-commerce). The related supply chain management practices became widely spread by the end of the 1990s. Popular perceptions of the potential effect of increasing ICT-use on logistics and freight distribution have assumed that the electronic transfer of information through an optimized logistics system would lead to more efficient transport operations – by substituting for apparently "unnecessary" transactions, by avoiding redundant traffic flows and by eliminating underutilized infrastructure. Electronic exchange would bring a new transparency into the logistics market that allows for optimal organization and allocation of transport services, and in some cases certain layers of the system (such as wholesalers, middlemen) would even become obsolete. However, recent experience revealed that both passenger transport and freight distribution systems appeared to be too complex, embedded in market situations and societal contexts, as if there were rapid and fundamental changes to occur (see Bowersox et al. 2000; Mokhtarian 2000; Capineri and Leinbach 2004).

The limited extent of impacts driven by E-commerce refers to the differentiated structure of the supply chain organization. On one hand, there are "front end" operations, comprising the interface between customer and vendor and thus organizing the order process of goods or services. In this regard, the Internet functions as a main platform for information exchange, translating informations into market transactions. Particularly in the case of the so called business-to-business E-commerce or "B2B"-relationship, this procedure offers certain advantages to both sides: The supplier receives orders to a substantially high degree of reliability and, often, predictability. Business customers can accelerate the entire ordering process, receive delivery faster, are able to follow the status of their order on-line. Changes can be easily processed within a fraction of the time required before. Large firms experience a higher amount of market transparency (offers, prices) and are able to increase their purchasing power in order to achieve overall cost reduction. For the business-to-customer E-commerce or "B2C"-relationship, the degree of change appears more limited: what once was ordered by phone or fax is now being transmitted through the computer and processed via the Internet.

On the other hand, different from these front end-operations is the way E-commerce affects the "back end" of the supply-chain: the storage, warehousing and physical distribution processes. It appears that this field of transaction has undergone only minor changes as yet. Even the most prominent web-retailers either did not spend much attention on physical distribution (which then raised costs and caused serious logistical problems) or have called for logistics providers to manage the entire supply chain and delivery operations. In general, the back end-practices of delivery are often similar to the way goods were being shipped five years ago. However, new front-end logistics operations may also require new facilities for storage, warehousing or distribution of commodities, what *The Economist* has called "the physical Internet" (*The Economist*, 17th June 2006, 13). In this regard, the conjunction of front-end and back-end operations turns out to be quite material and is insofar relevant for the aim and the subject of this study.

According to findings by periodical research, a majority of all E-commerce-transactions still refers to business-related exchange (OECD 2000, Riehm et al. 2003). Indeed the ordering and supply system within inter-firm relationships comprises the bulk of B2B-E-commerce. The advantage provided by E-commerce for firms is that it offers allegedly perfect market information and the opportunity to re-allocate the vast purchasing power of companies. The related cost reduction potential was estimated by OECD to account for about 40 per cent of corporate expenses, due to the opportunity for a complete breakdown and cost efficient re-organization of the supply chain (OECD 2000). In the distribution economy, information exchange by E-commerce helped improve inventory management and order fulfilment. Consequently facilities and vehicles are used more efficiently and many processes are at least partially more predictable. However, these technologies also allow for a more demand-side, short-term oriented and thus unpredictable order behaviour that can cause problems as well. The most common e-commerce applications in distribution, either in the logistics industry or in manufacturing and retail, relate to:

- order processing and fulfilment,
- warehouse and inventory management,
- fleet management, vehicle routing,
- freight brokering (e-market places),
- electronic payment.

Up to now, it remains unclear how far these applications have in fact become implemented in freight and distribution practices. Information management and integrated processing of the entire supply chain are supposedly used to a large extent, particularly by large firms. Inventory management is increasingly enforced with the assistance of mobile computers (as is routing or scheduling), even if that may not be a direct outcome of E-commerce. Other innovations, such as ordering via online-freight brokers, seem to be less accepted. There is no evidence that a majority of shippers and carriers would join electronic market places to boost delivery efficiency, as it was once expected. The spot market for short term transport demand may offer cost advantages, but this is normally balanced against other factors, such as reliability and predictability. Many of the electronic market places (freight platforms) that were established in the mid or late 1990s no longer existed a few years later. It is unlikely that this mode of operation will gain significant importance, except in market niches. From this perspective, the specific potential of E-commerce to rationalize and improve freight operations appears to be limited. It is more likely that those E-commerce applications will be implemented that explicitly fit into the predominant mode of rationalization and competitiveness. As a consequence, the direct implications of B2B E-commerce on the distribution system might be still limited. However, in the long term, the indirect outcome of logistics practices may increase although this remains difficult to predict.

Business-to-customer (B2C) related E-commerce has gained significant public attention during the rise of the New Economy in the late 1990s, although it is significantly less important in terms of number and value of operations than regular interfirm exchanges are. Whereas a wider public has been familiar with the idea of consumer goods home delivery for a long time, the electronic ordering process is fairly new and had only become possible with the advent of the Internet. Among E-commerce products, those that were predominant were either well suited for mail-order and delivery (such as books or computers) or were ordered both frequently and in bulk (such as grocery retailing). It is no coincidence that among the most important B2C or on-line retailer companies were, for example, the on-line bookstore *Amazon.com* or the personal computer and devices retailer *Dell*. However, the extent to which B2C E-commerce contributes to major changes in logistics, wholesale or retail distribution processes depends upon the rate of E-commerce adoption by users and market participants, which still seem to be limited.

Impacts of ICT and E-commerce on Freight Distribution

How far may information exchange by E-commerce lead to corresponding changes in logistics? In specific cases, physical distribution may be eliminated by E-commerce, for example, once documents or music files can be downloaded from web pages, instead of being packaged and mailed to a shop or directly to the customer. Electronic ticketing by airlines may spare the delivery of the tickets to travel agent or customer. However, those transactions still constitute a minority of on-line retail. Once they are just replacing the transfer of information or documents, the net effect may only be minor savings. In most cases, innovation caused by on-line retail is confined to introducing a new front end for sales and marketing purposes; what traditionally the mail-order catalogue was, is now becoming the Internet.

The back end of B2C is probably less sophisticated and more ubiquitous. Physical distribution is still required for delivering most products or services (Currah 2002). In many events of E-commerce, this task is fulfilled by logistics service providers and by express and parcel services, due to the large quantity of small parcels and packages in E-commerce related goods. In taking this approach, on-line retail makes use of highly efficient distribution systems, although it is associated with certain material effects, particularly by triggering additional shipments which consequently rise the demand for facilities that are essential to organize the freight flows. In other cases, on-line retailers such as *Webvan* were operating their own DCs and delivery fleet, which was clearly too expensive which finally contributed to push the firm out of the business. Most attempts to implement home-delivery were too costly both from the supply and demand side of distribution.

A major impact on the back end of on-line supply-chains is the elimination of one of the members of the distribution channel, compared to the usual marketing process operated by manufacturer, wholesaler, retailer and customer. Although there is empirical evidence that this is usual practice, the opposite can also occur. Middlemen (such as wholesalers) can fulfil an important cost-reducing function in the entire chain and in practice they have not lost real influence and significance. "The Internet story highlights a little understood trend: Wholesale distribution intermediaries are growing, not shrinking. Sales through distribution of non-durable goods, which include everything from books to industrial supplies, have been expanding at about 1.4 times GDP growth. Many pundits have predicted disintermediation, but the data shows the complete opposite trend. In other words, reports of the death of the middleman have been greatly exaggerated!" (The Lehman-Brothers 1999, 5)

Assessing the net effect of E-commerce on logistics and freight distribution is limited by the broad variety of distribution practices, the weak empirical data in terms of case studies, and the behavior of the users which still is hard to predict (OECD/ECMT 2001). Also cumulative causation has to be taken into account, including indirect or second order effects of the interaction of new technologies with their social and economic environment. As Visser and Lanzendorf (2004) have put it:

> ... the analysis of the freight transport effects of B2C E-commerce indicates that the latter increases in tonne volume terms (due to a one-off positive effect on demand and economic growth), alters spatial patterns of freight transport (due to shifts in consumption patterns), enhances service requirements of consumers (with considerable freight transport effects), and stimulates advancements in logistics and transport technology, along with outsourcing and the decentralization (sub-urbanization) of distribution systems. (Visser and Lanzendorf 2004, 201)

Fundamentals of Material Space: The Distribution Centre (DC)

Logistics Network Design, Infrastructure and Freight Flows

The changes in market environments and market behaviour as mentioned above deploy certain challenges for the logistics and freight distribution business. As a consequence of the related corporate reorganization and changing technological requirements, new facilities are becoming necessary: first, speaking quantitatively, the growth of logistics services in general propels the demand for more distribution space. Since manufacturing and retail firms have been consolidating or even outsourcing their warehousing activities, additional capacity is needed to accommodate growth and consolidation effects. Secondly, in more qualitative terms, logistics companies demand for buffer and organization space, according to the greater magnitude and complexity of freight flows and with respect to the increasing flexibility which customers demand. Third, even supply chain management which is primarily based on electronic front end-operations requires physical infrastructure and distribution. As a consequence of changing operational requirements, a rising demand occurs for a new type of facility, different from the old warehouse: the distribution centre (DC). The DC is one of the key elements in the modern supply chain. It is also highly important in geographical terms, since the emergence of the DC is associated with changing locational dynamics and regional developments. It can be thus also considered emblematic for the complex, interwoven series of impacts that are triggered by logistics modernization.

The traditional arrangement of goods flow included the processing of raw materials to manufacturers with a storage function usually acting as a buffer. The flow continued via wholesaler and/or shipper to retailer, ending at the final customer. Delays were very common on all segments of this chain and accumulated as inventories in warehouses. There was a limited flow of information from the consumer to the supply chain.

> The act of warehousing exists because companies are unable to predict demand and prefer to provide a buffer for themselves that accommodates spikes and lulls in the sales process. E-commerce tools, which enable the instant sharing of data among trading partners, dramatically improve the ability to predict demand. Aggregate demand for traditional warehousing space should decline over time, as the enabling technology is widely adopted and implemented. Today's state-of-the-art warehouses feature high-cube space with clear heights of at least 30 feet. However, as the new technology enables continual movements of products in the supply chain, the need to stack inventory begins to diminish. Traditional storage space must start housing activities that involve more horizontal movement rather than vertical stacking. (Kirschbraun and Bomba 2000, 16)

Consequently, the procurement of goods is no longer ensured by storing commodities, but through the timely delivery of consignments in the exact quantity and quality to the point of sale or, at least, to the facility where they are being provided for delivery to the customer. Thus firms are allowed to eliminate one or more of the costly operations in the supply chain organization. Accordingly, the concentration of storage or warehousing in one facility, instead of several, is going to become one of the most important physical outcomes of supply chain management. This facility is increasingly being designed as a flow- and throughput-oriented distribution centre instead of a warehouse holding cost intensive large inventories.

The key elements of this strategy are new logistic networks. Their inherent logic of rationalization is based on a high degree of centralization. As a result of the coarsely meshed configuration of logistic networks and parallel to the internationalization of commodity flows, the distribution areas are expanding (De Ligt and Wever 1998) and the number of facilities for handling goods is decreasing. Distribution is increasingly planned and operated on the basis of largely stretched networks, due to the premise of cost reduction. This kind of "economies of scale-oriented" network building leads to a shift towards larger distribution centres, often serving significant trans-national catchments. The structure of networks has also adapted to fulfil the requirements of an integrated freight transport demand, which can take many forms and operate at different scales. This includes both direct networks that mainly comprise point-to-point deliveries as well as hub-and-spoke-networks that include different hierarchies of connection.

Point-to-point distribution is common when specialized and specific one-time orders have to be satisfied. This often creates less-than-full-load (LTL) as well as empty return problems. The logistical requirements of such a structure are minimal, but normally at the expense of efficiency. Corridor structures of distribution often link high density agglomerations with services such as the landbridge where container trains link seaboards. Traffic along the corridor can be loaded or unloaded at local and regional distribution centres. Hub-and-spoke networks have mainly emerged with air freight distribution and with high throughput distribution centers favored by parcel services (O'Kelly 1998; SRI International 2002). Such a structure is only possible if the hub has the capacity to handle large amounts of time-sensitive consignments. The logistical requirements of a hub-and-spoke structure are consequently extensive as efficiency is dominantly derived at the hub's terminal. Routing networks tend to use circular configurations where freight can be transshipped form one route to the other at specific hubs. Pendulum networks characterizing many container shipping services are relevant examples of relatively fixed routing distribution networks. Achieving flexible routing is a complex network strategy requiring a high level of logistical integration as routes and hubs are shifting depending on anticipated variations of the integrated freight transport demand.

In the context of supply chain management and logistics network design, logistics companies demand for a new type of facility, different from the old warehouse: the distribution centre (DC). A DC represents a "physical facility used to complete the process of product line adjustment in the exchange channel. Primary emphasis is placed upon product flow in contrast to storage." (Bowersox et al. 1968, 246). It is no longer needed for storage but for the efficient consolidation of the materials flow. Despite the predominant "flow orientation" of the modern economy, warehousing remains necessary in many events, particularly since it is hard to predict the demand for certain goods delivery. The more varied and differentiated the markets are becoming, the larger the market areas are. The more competition is increasing, the more important is a finely tuned goods flow, mediated by buffers between suppliers and receivers of commodities. As a result, DCs are becoming key components in the supply chain.

The emergence of DCs means that logistics functions are going to be concentrated in certain facilities at strategic locations. This does not mean the demise of national or regional distribution centres, depending on the functional and organizational structure that may still require a multi-tier distribution system with regional, national or international DCs. The functions provided in a modern DC comprise receiving, storage, pick operations, value added activities, shipping, return processing and information management (Strauss-Wieder 2001, 10). One of the major tasks carried out in a DC is the consolidation of incoming freight and its immediate shipping to final destination (also known as "cross-docking"). Storage is practised in certain commodity groups that may not be delivered within short term. Added value is being pursued in post-production/pre-distribution processes, including assembly and customization (labelling, assortment), packaging, ticketing or product return and repair. The size of DCs varies and depends on its role, the composition of the network, the size of the market area and the volume of transshipments (Hatton 1990). With the trend towards concentrated supply-chain functions and thus to a decreasing number of DCs, the average size of a facility is steadily increasing, simply following the law of economics of scale. Hence it is not surprising that large DCs can achieve a magnitude of 50,000, 75,000 sqm or even more. Whereas regional distribution centres can go beyond the threshold of 100,000 sqm, large-scale or nationally oriented facilities are likely to exceed even that. This property of modern DCs raises many conflicts in terms of land use planning, infrastructure provision and the environment. Such large facilities can hardly be placed in traditional "gateway"-regions and certainly not within core urban areas in general.

The changing functional profile of the new facilities implies new locational requirements. Following the modern imperatives of mobility and accessibility, distribution firms necessarily locate at those places that offer excellent transport conditions. Secondly, they need cheap land for their increasingly large facilities. Most firms are taking these two particular considerations into account once they are looking for a site. This is due both to flow and stock-keeping aspects, regarding the high amount of freight traffic generated by DCs and their extraordinary demand for space. Trade offs between inventory and transport costs appear to be predominant, since freight transport and land use are closely intertwined (Ericksson 2001; McKinnon 1988: 152; Ryan 1999). Once transport markets became deregulated, total costs could be lowered by centralized locations, at the expense of higher transport costs. Final location decisions are made with respect to the network composition (i.e. the number of DCs) and the size of the markets that have to be served (Daskin and Owen 1999), both dependent on the type of industry or product group. Evidently, locational assets are not provided for without public policy and governance, even in a globalized, apparently unbound market economy. Zoning, economic development incentives, infrastructure provision and last but not least a qualified workforce remain important location factors.

Consequently, the rising establishment of DCs both in terms of numbers and of size raises serious questions and concerns (for more details see Chapter 3). First, the often large-scale facilites are by nature space consuming and, on one hand, tend to generate a high amount of traffic, since their establishment explicitly aims at functional concentration. On the other hand, it is suggested that functional concentration helps improving the overall transport and energy impact; empirical evidence is scarce in this regard with punctual findings that may not easily be generalized (Kia et al. 2003; Matthews and Hendrickson 2003).

Second, new players emerge on the real estate market, highlighting land capitalization and competition, but disregarding urban planning and integration issues and thus changing the planning conditions (Hesse 2006). The speculative nature of development activity raises land consumption and contributes to urban sprawl. Distribution firms particularly apply to this, since the comprehensive "orchestration" of material flows requires not only new sites but also extensive infrastructure, to connect inter-related places. Such prevalent commodification of land attracts further growth and agglomeration.

Third, whereas infrastructure provision was once a predominantly public task (e.g. in countries such as Germany), it is increasingly becoming subject to private corporate decision making. As a consequence, policy goals become more difficult to achieve: competitive dynamics between firms and – particularly – between municipalities do not allow for setting standards, demanding for commitments etc. The more speculative the nature of development, the more contingent planning will be. Thus the accelerated locational dynamics of logistics and freight distribution in general, and DCs in particular, is of increasing importance for regional development.

Chapter 3
Geographies of Distribution

Logistics and Freight Distribution in the Context of Geographical Sciences

Up to recently, academic geography did not pay much attention to logistics and freight transportation, as the focus was mainly on passengers and individual mobility issues. Textbooks on urban or general transport geography, like those edited by Hanson (2005), Taaffe et al. (1996) or Hoyle and Knowles (1998), have raised more freight related questions than they did in earlier editions, particularly with regard to trade and ports. The latter is probably the one logistics subject that received most reference from academic geography. Other core spatial implications of distribution and logistics have been directly addressed in geography by few authors who developed an insight into wholesale activities and their geographical distribution (Glasmeier 1992; McKinnon 1983, 1988; Riemers 1998; Vance 1970). Following the nature of retailing as an originally distributive activity, geographic research on retail and consumption is of interest in the logistics context too. However, retail geography does not pay much attention to distribution changes (Marsden and Wrigley 1996), although the physical movement of goods appears to be one of the costliest parts of retail activities (Christopherson 2001).

Why have the material world of physical distribution and the respective locales been so poorly covered by geographical sciences? One answer may be given by looking at the evolution of science and the path-dependent construction of subjects to scientific inquiry. The two traditional disciplines for investigating logistics and physical distribution are business administration (economics) and also transportation engineering sciences. Both cover, to varying degrees, aspects of space and location. However, the spatial character of their subject had been left out of most of their analysis. In turn, economic and transport geography did not develop too large a focus on logistics – keeping in mind the broad geographical relevance of distribution. A certain amount of research covers the planning aspect of freight transport in the urban context, either from a transport engineering and planning perspective or emphasizing related urban problems (Chinitz 1960; Ogden 1991; Woudsma 2001). The fact that logistics, as a factor of geographical studies, has remained relatively unexplored for a long time has certainly to do with historical improvements of transport technologies and conditions and the associated decline of transport costs. Spatial differentiation was always more or less directly linked to transport conditions and costs and the opportunities or frictions to overcome the limitations given by them. In the apparently "flat" world of the modern age, transport and logistics were taken for granted. Accordingly, the traditional transport cost-based determination of geographical sciences perpetuated a simplified construction of spatial variation on the basis of features such as the size, density and distance of flows.

More recently, logistics and freight distribution have been considered being more relevant. Freight related work received significant attention with respect to long distance trade issues. With emerging global trade, production networks and distribution systems, particular emphasis was given to ports and related research covering many of these issues (e.g. Hoyle 1990, 1996; Hoyle and Pinder 1992; Nuhn 1999; Slack 1998). In this context, an increasing amount of work on intermodal freight transport and terminal issues appeared as well (van Klink and van den Berg 1998; Drewe and Janssen 1998). Generally, international trade increasingly contributes to the amount and the nature of physical distribution. Thus globalization is now discussed as having a major impact on goods exchange (Janelle and Beuthe 1997; McCray 1998; Pedersen 2000; Woudsma 1999). Still, fallacies are noted in globalization discourses within economic geography, undermining the assessment of the role of transportation. Within the large body of work referring to the globalization discourse or the impacts of internationalization and free trade agreements, transport is not seen as a major issue or is *de facto* taken for granted (Holmes 2000). Even classic trade theory neglects the role of transport and logistics (Dicken 1998, p. 74), particularly the fact that transport costs have a fundamental impact on the amount of trade and goods exchange, as traffic constraints and opportunities in general do. It can be argued that this perspective is mainly the result of a misinterpretation of role of the transport sector, freight and passengers alike, as a derived demand. Under such circumstances, transport is perceived as a residual consequence – derived – of other processes or a mere "space-shrinking" function (see Chapter 2).

However, the absence of logistics and freight distribution the globalization literature in studies on international trade, multinational corporations and the division of labor/production is striking. Consequently, Wrigley addressed the following question (2000, 292): "Whatever Happened to Distribution in the Globalization Debates?" The fact that Dicken (2003, 471ff.) has meanwhile included a chapter on the development of distribution industries in the 4th and 5th editions of *Global Shift* may be considered as an evidence of rekindling interest in the topic, even if his contribution is more related to the globalization of retail businesses rather than the role of the core distribution sector. As logistics as a whole is gaining importance in the networked economy, it has recently been emphasized more frequently, both in the European and in the Anglo-American context (Hesse and Rodrigue 2004b, 2006; Hall et al. 2006). A geographical analysis of logistics may provide substantial evidence in that respect.

Regarding the nature of goods flows as originally derived from economic exchange, a considerable amount of research has been conducted on the production or commodity chain approach. This concept was originally developed by Wallerstein and Hopkins in the context of world-system analysis and received a broader distribution by Gereffi and Korzeniewicz (1994). Originally developed as an idea by Hopkins and Wallerstein (1986) in the context of world-system analysis, this concept received a broader distribution in the aftermath of conference documentation (Gereffi and Korzeniewicz 1994). The concept of the commodity chain is dedicated to "… a network of labour and production processes whose end result is a finished commodity" (Hopkins and Wallerstein 1986, 159). Thus the commodity chain represents a conceptual structure which is the basis of the analysis and assessment of interlinked processes not just in the each separate sphere of production, distribution or consumption, but with regard to their interdependencies. In terms of geographical research, this concept marked a substantial progress in linking the micro and macro dimensions of production and consumption. Meanwhile, a lot of research has been conducted in the purpose of further developing this concept, particularly regarding apparel and clothing chains, to a large extent on the food system, particularly on cut flowers, fruit exports, on jewelry, furniture, on industrial environmental policy; eventually even on drugs (see e.g. Hughes 2000, also the recent overview of Jackson et al. 2006 with respect to the emerging food chain issue).

The commodity chain approach delivers a more comprehensive understanding of the worlds of distribution – despite some shortcomings in terms of material space and probably transport. When Leslie and Reimer (1999, 401ff.) call for "spatializing commodity chains", it means in their words to highlight explicit geographical relations of the chain, like spaces and sites of consumption (for instance consumer's home), gender aspects and more. Taylor et al. note that there is a certain "[d]issatisfaction with the explanatory purchase of the commodity chain concept" […] that "has prompted an interest in commodity circuits. This was borne out of a concern, particularly amongst human geographers, that the concept of a chain is too linear, too mechanistic and too focused on the simple metric of length as opposed to other issues such as complexity, transparency or regulation." (Taylor et al. 2006, 132)

Accordingly, in this particular distribution context, analysing chains should also include emphasizing the material dimensions of goods flow, the transport and logistics infrastructure in their significant spatial impacts and power relations as well. Analytically, the commodity chain approach needs to be linked to the issue of site and location, since it concerns distribution not only as a basic organizational infrastructure of the entire network economy, but then also mirrors the spatial arrangements of the interrelated spheres of production and consumption. In this case, it offers a promising perspective for further investigating and assessing the geographical extent and significance of logistics and freight distribution.

Toward New Geographies of Distribution

Geography is originally about space: the investigation of geographies of distribution primarily aims to analyze the issue of location, particularly the location of facilities that are dedicated to the handling and transhipment of consignments. Traditionally the locations of freight distributors were dependent on industry: distribution sites were located at both ends of the production-oriented transport chains, at the place of production (manufacturing location) and at the place of either consumption or transport to the end consumer (retail trade). These were generally urban agglomerations from which the distribution areas were supplied. Depending on the focus of the functionally segmented transport chains, they chose proximity to either the production site or to the market. On the whole the distribution system had a strong affinity with the central place system of settlement and market areas. The decentralization of industrial production and the parallel expansion of transit routes logically led to different choices of location for handling and storing freight (Pred 1977). This was stated by Chinitz (1960) for the greater New York City/New Jersey area, who has put it as follows:

> Wholesale establishments, warehouses and terminals for both water and land transport – all of which are characterized by continual in-and-out freight movements – have been increasing their employment fastest in places outside the congested heart of the Region. And in this respect they closely resemble manufacturing itself. (Chinitz 1960, 153)

Recently, technical and organizational changes in distribution have transformed location systems further. "Much of the goods production, transportation, and distribution jobs that core cities have depended upon will continue to decentralize to outer suburban and ex-urban areas and to lower-cost, smaller and mide-size metropolitan areas." (U.S. Congress Office of Technology Assessment 1995, 145) The crucial factors here are: first, the transition from a single plant or site to networked production (Schamp 2000) and, second, the continuous improvement of transport and communications technologies (Barry and Slater 2001). It would otherwise have been impossible to streamline materials management; and the rising locational dynamics of logistics is an outcome of policy and politics: After a period of intense liberalization of the highly regulated transport markets, the relationship between fixed and variable costs for corporations has significantly changed. The respective trade-offs between rising capital and inventory costs on one hand and falling transport costs on the other hand were highly supportive for a new locational mobility. Consequently, logistics networks became redesigned, some locations were closed and others expanded or newly established. The related concentration effects allow the utilization of economies of scale and thus contribute to making the business more profitable.

Weakening Ties at Gateway-locations

Among the location complexes most affected by changes in the logistic requirements of enterprises are the traditional gateways of freight traffic – especially seaports, container ports and large freight airports in industrialized countries, but also increasingly in areas of neo-industrialization where globalized transports are concentrated. Port locations are now confronted with pressures arising from the restructuring of the value-added chains. As nodes of long-distance transport they do not act as traditional gateways for supplies to their hinterlands (Nuhn 1999, 88); rather, they provide the organization of the physical freight flow in a larger spatial-organizational context. In this way they are integrated in the space/time organization of complex transport and production systems. Increasingly value is being added on the high seas or inland, i.e. away from the ports. Hence individual locations are becoming less important. New transport and container technologies, automatization of many work processes and flow orientation in the logistic creation of added value mean that these activities are no longer necessarily tied to port locations (Läpple 1995).

All big seaports are now keenly competing to operate as international hubs for logistics services (van Klink and van den Berg 1998; Notteboom 2004). This applies to the big ports such as Rotterdam, Antwerp and Hamburg on the North Sea coast, New York/New Jersey on the east coast and Los Angeles/Long Beach, Oakland and Seattle-Tacoma on the west coast of the USA (McCalla et al. 2004). In Southern California, the Alameda Corridor was built in Los Angeles/Long Beach to improve the links between the ports and their hinterland, e.g. the quickly growing industrial areas in San Bernadino County in the so called "Inland Empire". Since significant proportions of industrial production have been relocated to the Far East, new competition has also arisen in the Asia-Pacific Region: the new hubs of international freight flows have evolved in Hong Kong, Singapore, Shanghai, Pusan or Kaohsiung (Comtois and Rimmer 1997). Being faced with rising competitive pressure, all locations are committed to expand their infrastructure like enlarging port or transhipment areas, dredging rivers etc. On the European continent, the "Betuwelinie" is dedicated to provide a rail link between Rotterdam Port and the Ruhr area – possibly also the further hinterland. The A73 motorway between Venlo and Nijmegen has proved a successful logistical corridor for relocating firms ("de logistieke snelweg"). The Betuwelinie is intended to alleviate some of the A73's congestion. A similar target is envisaged for the planned rail corridor between Antwerp and the Ruhr area, the "Iron Rhine". Yet, both the Betuweline and the Iron Rhine suffer from a cost-benefit ratio that justifies the significant investments that are required for operation at a competitive level.

Like the distribution networks of the transport industry in general, these strategies are governed by the economies of scale that favour the concentration of the largest possible volume of freight (Slack et al. 2002). However, these efforts are challenged by significant constraints almost everywhere: Space is scarce and expensive in traditional port locations. Scope for expansion in the densely built core cities is either rare or politically controversial. In addition, traffic density has increased as a result of higher volumes of freight transport. "Modern ports bring both prosperity and problems. They bring trade and wealth, but at the same time they bring crushing, uncontrollable congestion of road and rail; costly construction and maintenance for landside infrastructure; enormous costs associated with dredging and underwater channels and disposing of the dredged sediments; hot competition for an urban area's most valuable land – on the waterfront; and an almost exquisite vulnerability to terrorist attack via the very containers that form the basis of modern trade and wealth." (Weisbrod 2004, 47) Hence, various alternative concepts are being developed at these locations as well as the classical expansion strategies (space, navigation channels, etc.). These include reclaiming land from the sea (e.g. the Maasvlakte in Rotterdam), building sea-based ports off-shore (Weisbrod 2004) or creating brand-new coastal ports like the planned deep-sea Jade Weser Port in Wilhelmshaven, Germany. For quite a long time distribution and logistic facilities have been expanding into the hinterland. Allaert termed this process "sub-harbourization", a analogy with the suburbanization of distribution (1999, 3). In the case of large port regions, Notteboom and Rodrigue (2005) also speak of "port regionalization", a displacement of port functions into the region. Notteboom and Winkelmans (2005) described these spatial polarization and expansion movements with reference to the Benelux seaport system (cf. Map 3.1).

Van Klink (2002, 143ff.) relates this particular development to a third out of four stages of modern port development (see Table 3.1). Whereas the first and second stages consist of the change from cityport (or: "port city") to "port area", including the expansion of port-related land and infrastructure beyond city limits or at least beyond the traditional vicinity of the port, the third stage contains a development of a certain "port region". This port region is characterized by the influence of globalization (unlike the two former phases, which were shaped by trade and industrialization): it adds more functions to the port area, particularly container distribution, focuses predominantly on container flows instead of bulk

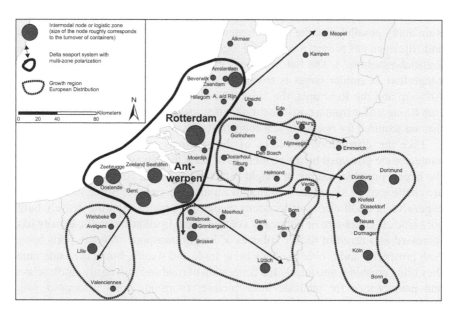

Map 3.1   Polarized seaport system

Source: Own.

Table 3.1   Stages in the development of seaports

	Stage 1	Stage 2	Stage 3	Stage 4
Fundamental development	Rise of trade	Industrialization	Globalization	Informatization
Main functions of the port	Cargo handling Storage Trade	Cargo handling Storage Trade Industrial manufacturing	Cargo handling Storage Trade Industrial manufacturing Container distribution	Cargo handling Storage Trade Industrial manufacturing Container distribution Logistics control
Dominant cargo flow	General cargo	Bulk cargo	Containers	Containers + data
Spatial scale	Port city	Port area	Port region	Port network
Role of port authority	Nautical services	Nautical services Land and infrastructure	Nautical services Land and infrastructure Port marketing	Nautical services Land and infrastructure Port marketing Network management

Source: van Klink 2002, 145.

or general cargo and needs further land reserves for operation and expansion. Van Klink analyzes such developments by comparing the ports of Antwerp (Belgium) and Rotterdam (Netherlands): the two major competitors of Hamburg in the North Range. Both ports in their historical trajectory reveal a significant shift from original port areas toward hinterland and region. The suburbs were chosen for several reasons, one of them was the high labor cost of dockworkers, which made labor markets in more distant municipalities more attractive. Also, land costs were lower outside ports, and locations along the highways offered better accessibility to the hinterland than the ports themselves (van Klink 2002, 153).

Re-locating facilities and infrastructure even farther into the hinterland is now becoming an increasingly popular alternative way of coping with congested mainport-areas. As a consequence, "inland hubs" are emerging, where primarily road and air freight are consolidated. These new DC areas are mainly connected to the interstate network and air-cargo facilities. As a result warehousing, trucking, freight forwarding and air cargo activities also became major indicators and drivers of this new distribution economy. One of such new inland hubs is emerging along the Ohio River Valley, following a corridor from Ohio and Indiana to Tennessee.

The "first generation" e-fulfilment providers are gravitating towards the preferred location for a single, centralized distribution facility, the greater Ohio River Valley, namely the states of Ohio, Indiana, Kentucky, and Tennessee. Industrial markets such as Columbus/OH, Indianapolis/IN, Hebron/KY (Cincinnati/OH) and Louisville/KY have seen substantial demand from these users. (Abbey et al. 2001, 15).

In 1997, more than 150 distribution centres larger than 50,000 square feet, were located only in the City of Columbus/OH. Both inventory and recent absorption in the Columbus industrial real estate submarket belong to warehousing to 80 per cent (SIOR database. 2001; see also Urban Land Institute 2004).

Concomitantly, European Distribution Centres (EDC) are becoming larger, as the pressure to consolidate distribution centres into pan-European centres continues. With access to a significant part of the European market place required, core Europe is the preferred location – most notably Benelux and eastern France. National and regional centres are under pressure in all these countries as distributors attempt to offload this layer of warehousing. The Netherlands is famous for being among the most favored location for European logistics, due to excellent accessibility, advanced terminal and transport infrastructure, critical mass of logistics functions and attractive operating conditions (vis-à-vis its neighbours). Schiphol Airport and the Port of Rotterdam are among the most important hubs for international freight flows in Europe. Major population concentrations are well represented in the following areas: Paris, London, the Ruhr area and Frankfurt (Europe's largest air-cargo hub). Flanders in northern Belgium and the Nord-Pas de Calais region in northern France also score highly. Due to improved access to the continent via the Channel Tunnel, UK distributors tend to prefer north-west Europe (cf. JonesLangLaSalle 2001).

Spatial Shift at the Metropolitan Regional Level

The growing importance of the distribution and logistics sectors is also affecting agglomerations and other densely populated areas. One reason for this is the tendency of value-adding and logistics chains to expand from the big intersection locations into their hinterlands; another one is the proximity of agglomerations to customers. In the course of the economy's growing demand side-orientation, distribution locations are moving away from production and towards consumption, i.e. partly towards the agglomerations again (Mueller and Laposa 1994). This is where clusters of distribution centres form, sometimes at single, more or less unconnected locations, sometimes planned as freight transport centres (Hesse 2004). The locational advantage of agglomerations is less their position in an important infrastructure intersection, but rather their combination of short- and long-distance accessibility and also access to major distribution areas. Decisions on the location of new DCs are primarily based on the criteria of size and accessibility. In the past few decades, this combination of factors has brought a greater proportion of distribution uses to the areas surrounding agglomerations, as industry already did before. Considering the present conditions of flow-oriented economy, this movement out of the cities has become stronger because the core cities and their traffic congestion create more and more obstacles to flow-oriented distribution.

Glasmeier and Kibler (1996) also attribute this development to new technologies and their application in transport and transhipment; a different arrangement of the supply chain and a different power structure in the logistic channel always trigger changes in the spatial organization of distribution too. However, this does not apply equally to all branches of logistics: the same authors (ibid.) suggest that wholesaling will still remain in core or even central city locations, as Vance (1970, 130ff.) already stated much earlier, but Glasmeier and Kibler (1996) assume that the de-centralization of logistics will continue:

> We suggest there will be a variety of countervailing locational tendencies, including the decentralization of large mass wholesalers and their warehousing complexes, combined with increasing levels of specialization of the remaining inner-urban wholesale and warehouse functions. … We anticipate a general reduction in wholesale employment, particularly in inner-urban areas as large wholesalers move their warehousing operations out to rural and adjacent suburban areas to take advantage of cheaper land and labor. (Glasmeier and Kibler 1996, 741)

Empirical studies on the extent of this trend are rare, however. McKinnon studied the development of warehousing in England (McKinnon 1983) and traced its spatial distribution in industry and distribution services. He revealed a spatial pattern of sub-areas around conurbations and along major motorway corridors with above-average growth rates (McKinnon 1983, 392). This applies especially to the northern part of Greater London (M1–A1), whose growth is attributed to the fact that local distribution moved out of the core city. These locations are also attractive for regionally scattered supply networks (see Table 3.2).

Riemers (1998) employed a similar approach in her study of the Dutch wholesale sector with reference to the changing spatial distribution of the working population between 1973 and 1993. Employment in the wholesaling sector declined in the three biggest Randstad cities (Amsterdam, Rotterdam, The Hague), whereas four regions in the southern Netherlands (Brabant) recorded higher figures (Riemers 1998, 90). These regions are favourably located in the heart of the Benelux region with ideal connections offered by the motorway network. Locations in the surrounding areas profit from the fact that local and regional planning actively supports the relocation of enterprises by designating "city distribution centres" on the outskirts of cities and, at the same time, restricting truck traffic in city centres (ibid.). By contrast, peripheral parts in the northern and eastern area of the Netherlands, for example, have lost out:

> The general tendency seems to be that wholesale businesses look for more central locations in the open space within the Randstad or to the eastern and southern fringes of the Randstad. (Riemers 1998, 90)

Table 3.2   The modern place of goods handling in urban and suburban areas

	Function	Location	Examples
"New" centres of distribution at the urban periphery	Spatial anchor or magnet of modern logistics and distribution networks (second outward drift)	Areas at motorway intersections with cheap land and workforce, close to the customers" area (urban markets)	Shopping malls," big box" commercial areas, industrial DCs and warehouses, almost ubiquitious
Large scale distribution of / for retail, wholesale, warehousing	Decoupling of distribution from the urban market place (counter-urbanization related drift)	Peripheral regions with cheap land, workforce and motorway access	National HUBs of distribution firms, pan-European DCs, inland-ports e.g. in the Ohio-River Valley
Interregional Mainports	Gateways of the global and international goods flow	Selected sea ports, large freight airports,	The ports of Los Angeles/Long Beach, Rotterdam, Hamburg, new airfreight hubs in the U.S. Mid-West

Source: Compiled by author

The locational structures of logistic services show a similar spatial logic in Flanders (Belgium) to the north of Paris (Jones Lang LaSalle 2001), around London (McKinnon 1983) or Milan (Debernardi, Gualini 1999). In Germany, too, terminals, depots, and new distribution centres are increasingly being built at the periphery of urban regions. In eastern Germany there are the corridors along BAB 14 (Halle-Leipzig), around Hermsdorfer Kreuz in Thuringia (A 4/A 9) or along the Berlin Ring (A 10). Typical examples in western Germany include Hannover (cf. Hannover Region 2000), the eastern Ruhr area, the Lower Rhine, eastern Munich and the region around the Frankfurt/Main airport. Employment statistics in Germany's transport sector reflect this trend: in the late 1990s, such jobs were concentrated (in absolute values) in agglomerations and in the port districts of Hamburg and Bremen. However, districts such as Groß-Gerau (near Frankfurt/M.), Saalkreis (near Halle/Saale) and Unna (North Rhine-Westphalia) had the highest density or the highest relative proportions of this sector in employment figures as a whole (Bertram 2001).

"Filling in" in Peripheral Regions

The advent of national or even European transport networks and the poor state of infrastructure in core urban agglomerations means that even places outside the agglomerations have become relevant as potential locations for logistical functions. There are two reasons behind: first, the traditional disadvantages of poor accessibility have been offset by extensive motorway construction during the past few decades; travel time to these markets has dropped considerably, despite their peripheral situation. This makes such locations attractive, especially for supra-regional networks. Second, these regions possess space and labour-market reserves; at least the former factor offers an advantage over agglomerations. Hence, previously near-border areas are well represented in this logistical logic of location: because of the demise of national borders these areas enjoy new locational advantages, they are easily accessible by motorway and have huge reserves of space. The prototype of this is northern and central Hesse, Germany, whose central location since 1990 has made it a suitable node for national transport networks. Especially the area around Bad Hersfeld now counts as a kind of hub for nationally oriented distribution centres (Gudehus 2000), as the main terminals for many parcel services, the German headquarters of an internet dealer, and the distribution centres of retail and wholesale companies are based here.

In western Europe the region around Venlo in the Netherlands is increasingly filling the earlier vacant space between the Randstad and the Ruhr area. It is gaining a reputation as a hub providing logistics service for the agglomerations and throughout Europe. On the whole the Benelux countries are on a development streak owing to their easy-to-reach location (De Ligt and Wever 1998); the same holds true of the Nord-Pas de Calais region in northern France, which has acted as a bridgehead between Great Britain and continental Europe since the Channel Tunnnel was built. Cabus and Vanhaverbeke (2003) have studied similar trends in western Flanders (Belgium). Its rural areas are under strong suburbanization pressure from the core area of Belgium, and they are experiencing a highly dynamic development. Business services show the highest job growth rates – for instance in the transportation and logistics sectors:

> [...] there are specific niche economic activities performing very well in the countryside. This is the case for business services, transportation and catering. It is obvious that the excellent growth figures for transportation and logistics are linked to the proximity of world harbours, a favourable situation regarding traffic congestion, and the availability of cheaper space – an important requirement for the logistic handling of goods. [...] As a result, new logistics developments such as European distribution platforms (EDP), become more important and play a crucial role in the functioning of production systems. (Cabus and Vanhaverbeke 2003, 241)

The peripheral regions are showing clear signs of the ambivalence of logistical modernization in terms of cost-benefit ratio. On the one hand, rural regions have lost economic substance through the breakdown of regional production chains (Nuhn et al. 1999). Moves towards centralising production and distribution have caused large-scale collapse of the traditional industrial pattern of peripheral regions, e.g. the processing of agricultural products. At the same time, rural regions are attractive target areas for relocations from the agglomerations (Bade and Niebuhr 1999). In this respect, these regions can also benefit when logistics and distribution move to the periphery. It is sometimes assumed that logistic nodes and locations are attractive sites for production facilities because of their transport connections; however, there is yet no empirical evidence for such links. Not in the least because almost all these locations are equally accessible. In the case of certain rural areas, logistic projects are increasingly receiving active economic support because these regions were often not traditional industrial locations or have lost their industrial base and seek for substitutes in terms of growth and employment.

Institutional Change and Spatial Shift in Logistics and Freight Distribution

Institutional Change on the Demand Side (Logistics and Distribution Firms)

Institutional change, particularly the opening of transnational goods and service markets and the deregulation of the transport markets supported by the advances of new technology, had a significant impact on logistics networks and operations. First, it contributed to the emergence of the globally operating logistics corporation. As a consequence both – logistical networks and nodal functions – are changing as well (Notteboom 2004). The pressure of competition is particularly threatening the traditional hubs: they are no longer competive mainly for size, location or infrastructure reasons (see above), but have to run against each other for attracting the transhipment of consignments on a highly contested market. In terms of policy and administration, they are also increasingly run by multinational service corporations that operate a global infrastructure network, instead of local port companies or authorities based on municipal bodies (Nuhn 2005; Slack and Frémont 2005).

Owing to the global activities of these corporations, the functions and services usually concentrated at major hubs such as seaports or airports are losing their traditional locational ties; and infrastructures and locational systems are becoming increasingly mobile. Marginal locations can suddenly achieve great importance merely because of the individual strategies of globally operating services. One example is the container port of Gioia Tauro in Calabria (Italy) that had become one of the major European terminal complexes within a ten-year period. This is where part of the container freight coming through the Mediterranean from the Suez Canal is reloaded onto the railway and transported to northern and central Europe. The volume of goods handled at Gioia Tauro already totals almost 50 per cent of of what is handled at the Port of Hamburg (Notteboom 2004, 99). On the Arabian Peninsula, Dubai is now a boom location; as a global node, it is evidently positioned ideally in time and space. A *Logistics City* is under construction in the immediate vicinity of Dubai's existing container port and a free-trade zone; ultimately it will cover 25 times the area of *Cargo City Süd* in Frankfurt/Main. There are also plans for a huge freight airport (Helmke 2005). The German freight-forwarding firm Schenker now operates a global air-freight link between Dubai, Toledo (Ohio, USA) and the secondary airport Hahn in Germany, thus offering global services between the Middle East, Europe and North America within a time span of less than 15 hours. The case of Dubai perfectly demonstrates how first the restructuring of global supply flows creates complex spatial arrangements and second triggers economic development at places that were formerly considered remote and peripheral. In line with that, institutional reforms are shaping port authorities that were traditionally in the responsibility of public agencies. The now privatized port authority behaves in the freight market in a way that Hayut (1981, 160) had already coined as a certain "aggressiveness of port management".

The locational shift and dynamics of freight distribution at the various spatial scales is not at least an outcome of high pressure on supply chains, caused by accelerated information transfers, changing consumer preferences and rising competition among logistics businesses. Since many parts of the supply chain are now globally integrated, facilities such as distribution centres tend to be the link between global sourcing and regional distribution. As a consequence, the DC has become an interface between the geographies of manufacturing and retailing, handling the distribution scale and scope. Innovations such as containerization and particularly IT developments have integrated all components of the chain. In response, major players in the distribution business (e.g. container shipping lines, freight forwarders, warehousing firms, terminal operators) are trying to control as many parts of the logistics chain as possible. Not coincidentally these firms are challenged by vertical and horizontal linkages, by mergers, takeovers and strategic alliances (Slack et al. 2002). Judging from their particular perspective, staying competitive means to increase the throughput and provide the demanded services at low rates. As a result, the activity space of main ports is increasingly becoming relocated to low cost locations reaching far beyond traditional terminal sites and connecting more distant places of their hinterlands. It seems to be likely that the role of (large) firms cannot be underestimated in this respect:

Especially the larger logistics company seem to control more and more the actual routing of transport flows. They are able to decide in relativ autonomy through which (main) port flows run, where warehousing takes place and what modes of transport are used. (Meijer and Ten Velden 1996, 49)

Corporate decision makers, particularly those of large firms, have the power not only to orchestrate large bundles of flows, but to "produce" space by the configuration of networks, defining related nodes and selecting certain locations, since they develop places according to their needs. In such locales, the different scales and layers of the networks are becoming assembled and materialized. Therefore, as a consequence of globalized production systems, logistics and freight distribution is not only required to complement the manufacturing process and to carry the final product to the market. It is the functional and geographical integration of the global economy that is based on core components of the distribution network: flows (information, money, commodities, vehicles), nodes (ports, airports, railyards) and – all interlinked – networks. It is, in fact, the global distribution system, consisting of firms, transport modes and infrastructure, which makes the global economy running. With regard to the significance of logistics, the location of economic activity is no longer a mere function of transport infrastructure provision, but also relates to the ability of regions and cities to cope with the extraordinary demand for flexible, timely and cost-efficient physical distribution. As a result, the intertwined processes of globalization, the emergence of GPNs and new corporate strategies articulate a specific demand for appropriate infrastructure. Places that accommodate such demand may improve their competitive position significantly – both in terms of regional economics in general and with respect to locating the transport industries. Hence, against the background of more recent trends in technology and with respect to globalization, the case of regional development of logistics and freight distribution becomes more and more important.

Institutional Change on the Supply Side (Real Estate)

Institutional change also occurs on the supply side, particularly with respect to those actors that provide distribution firms with space – notably land – for placing DCs. Traditionally, logistics real estate was a mere subsidiary of commercial or industrial real estate markets, different from retail or office markets, but not to a large extent. Following the rise of the logistics, the real estate industry is now specializing, in order to better meet the demand of distribution businesses. Besides market growth and changing user requirements, this diversification was primarily based on two issues. First, this subsector is characterized by a high market capitalization: prime yields reach a level of eight per cent per annum, which compares very well with retail investments. Consequently, capital markets develop a rising interest in logistics real estate. A second issue that raised an increasing interest in logistics real estate was associated with the expected explosion of electronic commerce (E-commerce) in the late 1990s (see above). With the advent of the Internet as a universal communication mode, online merchandise appeared to be a source of accelerated growth in trade and transactions. Since most of the electronically traded goods require physical delivery, logistics received much interest as a tool for online-retail and wholesale. In this respect, logistics was considered a "backbone" of the New Economy without which any of the new businesses couldn't successfully operate. More recently, this prediction came true when E-commerce firms went bankrupt, due to their neglect of basic distribution expertise, cost and requirements.

Not surprisingly, an increasing number of studies on the emerging logistics real estate market emphasized the significance of the new market segment (JonesLangLaSalle 2001a, 2006; DEGI Research 2006). This is particularly based on the generic growth of the logistics market, new geographical patterns and locational strategies as an outcome of logistics consolidation, and on the fact that demand for qualitative space creates the need for further development which is certainly interesting for real estate firms. The market reflects changing behaviour patterns on both demand and supply sides: The demand side consists of distribution firms in wholesale, retail and the transport industry. Their attitude to real estate is characterized by changing purposes (e.g. contract-related instead of firm-related), by a changing timeline for use (short-term cycles instead of long-term) and by a changing market behaviour (lease or rent instead of ownership). Due to the rise of contract logistics as an outcome of outsourcing and periodical tendering by manufacturers or retailers, distribution firms are increasingly making flexible location decisions: The company is no longer committed to life-long location but follows the ever-changing pattern of contractional flows. Thus locations are becoming adjusted to the temporary nature of contracts. Related operation and planning horizons appear much shorter than before, comprising three to five years (or even less) instead of up to ten years before. This also implies a different purchasing behaviour, because firms like that are less inclined to buy land. As a result, rent and lease is being favoured by European logistics firms.

Significant shifts on the supply side are associated with the emergence of specialized developers and real estate-brokers. Since sites and facilities are no longer user-owned, there is a bigger role for intermediaries such as development and brokering firms. Brokers trade sites in order to find customers, or they are asked by customers to seek appropriate locations. Developers purchase, own, develop and rent out land for distribution and related purposes, and they also trade facilities. Both the brokers" and developers" activities emerged out of their general commercial and industrial experience, which is now being directed towards the particular demand of distribution firms – either by opening up specialized branches or by founding new firms. Many of them are global often owned by major U.S. corporations. Also, investment trusts and banks are increasingly becoming active in the logistics sector, in order to provide resources by capital funds. With the outsourcing of services that is now being practised in the process of site selection, development and management as well, a new phase in the commodification of land is becoming visible.

In order to evaluate the consequences of these changes: What does the emergence of brokers, developers and investment funds in the distribution business mean for regional development? First, according to an institutional view on property development (Guy and Henneberry 2002), the changing institutional framework is likely to shape the whole way land devoted to logistics is becoming a commodity (see Figure 3.1). This characteristic is by no means new. However, potential distribution sites are now being assessed and traded in a completely new way. From the developers' perspective, the first and foremost goal is to achieve a revenue on the invested money instead of providing the conventional "public goals" related to freight transport. In this respect it makes a difference whether logistics firms, e.g. SMEs, are going to build facilities for their own use and at their own risk, or whether investment companies are going to supply an anonymous market, following more or less speculative purposes. The performance of the U.S. warehousing market is in this regard different from Europe. It is characterized by a higher degree of rent or lease (instead of ownership), by intermediate activities

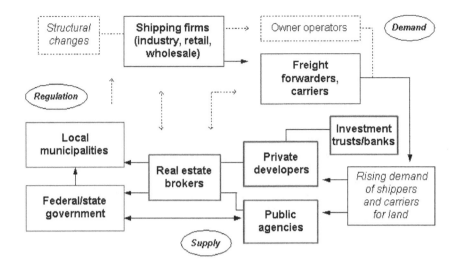

Figure 3.1   System of actors in logistics real estate development
Source: Own.

and speculative development of land. The consequence: More speculation is likely to increase the consumption of open space.

Second, from the service providers' perspective, the criteria for site selection are increasingly becoming multifaceted, particularly with the rise of large-scale network configurations as the overarching structure for corporate logistics strategies. Besides specific product and customer related requirements, scale and composition of the distribution network appear as decisive for the question where to locate a DC. This is particularly true for the upcoming trans-European distribution networks that are designated to serve the populated core of Eastern or Western Europe (Hoppe and Conzen 2002). According to the concentration in the distribution industry most of the network related site-selection is based on the active participation of brokers and developers. Once the spatial scale increases such commodification of land leads to a certain abstraction from the concrete place, in favour of the network structure. As a consequence, the network tends to determinate the node as a single local entity. This also means that local criteria, e.g. the adaptability of the node, are being outweighed against the overall network structure.

The rising significance of intermediaries is important for the way land is politically regulated, especially locally. Regarding the recent trend towards privatization of infrastructure provision, this type of land development raises serious conflicts between private and public goods and interests. Market capitalization and returns on investments are now becoming preferred in land use decisions, whereas public institutions – obligated by environmental, transport or community needs

– are losing influence. As a result, a power shift in land use conflicts occurs, likely to exert an increasing pressure on public policy to open up land for development.

Accordingly, the newly emerging patterns of land demand and supply are going to affect regions in two ways: First, new players emerge on the real estate market highlighting land capitalization and competition, but disregarding urban planning and integration issues. The speculative nature of development activity raises land consumption and contributes to urban sprawl. Distribution firms in particular apply to this, since the comprehensive "orchestration" of material flows requires not only new sites but also extensive infrastructure to connect inter-related places. Such prevalent commodification of land attracts further growth and agglomeration. Thus distribution takes over the classic role of industry in shaping the territorial organization: it "produces" places (Storper and Walker 1989, 70). As a result, "regional complexes of distribution" emerge, places dedicated to the handling of goods, either being isolated (DC) or, whenever that is the case, more integrated (IFC).

Second, whereas infrastructure provision was once a predominantly public task (at least in Germany), it is increasingly becoming subject to private corporate decision making. As a consequence, policy goals become more difficult to achieve: competitive dynamics between firms and – particularly – between municipalities do not allow for setting standards, demanding for commitments etc. (see Table 3.3). The more speculative the nature of development, the more contingent planning will be. Third, even ambitious public agency-plans do not necessarily ensure the achievement of public goals. This is due to the market imperative of acquiring firms and selling land – goals that must be respected by public development bodies as well at the risk of their own failure. Aims other than land development like promoting intermodal freight transport may not be pursued to the same extent.

Table 3.3   **Public and private development of logistics real estate in comparison**

	Integrated freight centre	Private logistics real estate
Development	Policy oriented (transport, economic development)	Capital oriented
Occupier firms	Logistics, distribution, related services, others	Logistics, distribution
Major players	Public agency (State-based), local municipality	Private developer (international), real estate agent
Contracts	Lots for sale	Rent or lease
Location	Partly integrated	Often isolated
Traffic access	Multimodal (road, rail, partly water)	Road traffic

Source: Own research.

Speaking in general terms of policy and planning, it appears that the power relations between public and private actors are further shifting towards the private.

Cities and regions are particularly affected by these changes, since the geographies of distribution are based upon newly emerging nodes within large-scale networks (Amin and Thrift 1992; Smith 2001). These nodes are physically grounded in urban and metropolitan places, representing a certain "spatial fix" of logistics. Hoewever, the local "embeddedness" of modern sites tends to be limited, since firms try to get rid of traffic jams, to bypass the rigidities of planning requirements or to minimize the power of labour unions – factors that are more prevalent within urban regions than at their periphery. Therefore changes in distribution have material implications for planning and policy. The most urgent need is establishing a general awareness of the distribution economy. Furthermore, public sector actors should focus i) on incorporating the sector of logistics and freight distribution into local and regional planning, ii) the case of the intermodal warehouse or DC deserves more consideration. Implementing such policy goals is the joint responsibility of private and public actors, the more dominant private interests may appear (see below).

Chapter 4
The Berlin-Brandenburg Case Study

The Berlin-Brandenburg Metropolitan Area, Germany

The Berlin-Brandenburg region comprises an area of about 5,400 square kilometres with about 4.2 million inhabitants and about 1.5 million employees. Of that, 3.4 million inhabitants and 1.2 million employees are concentrated on a territory of about 880 square kilometres belonging to the City of Berlin, the rest makes up the suburban area which is part of the State of Brandenburg, also called the "Zone of Mutual Interdependence" (see Table 4.1). The region was split into two parts by the post war partition of Germany that lasted from 1961 until November 1989, the time when the wall came down and Germany became unified afterwards. Today the settlement structure is characterized by the sharp contrast between the densely populated core city and the scattered, dispersed suburbs. In between, there are particular areas that are neither city nor suburb: they are located within city-limits, yet they appear more suburban in terms of density and urban design.

These areas, to a significant extent occupied by single-family homes, have represented the typical target-space for suburbanization for decades. For political reasons, the move to the suburbs in the former Berlin (West) had to remain within city-limits. In contrast the centrally planned urban politics in the former East-Berlin, then Capital of the GDR, pursued the development of a "Stadt-Umland-

Table 4.1   Spatial development in Berlin and suburbs, 1990–2004

	Berlin	Suburbs*
Area in skm	891	4,448
Inhabitants 2004	3,387,282	992,200
Change 1990–2004 in %	-1,3	26,4
Inhabitants 2015 (Suburbs 2020)	3,362,0	1,015,2
Change 2001–2015/2020 in %	-0,8	5,0
Population Density (inhab./skm)	3,799	224
Occupation (place of work) 2002	1,109,610	279,5
Change 2002 zu 1996 in %	-11,6	-10,5
Unemployed 12/2002 (Berlin 6/2002)	290,544	73,965
Change 1993–2002 in %	29,8	53,5

* Brandenburg area of the Zone of Mutual Interaction (eV)

Source: own after GL 2006

Region" (Heidenreich 1972), which however carried only a few features of suburbanization. Both in the former eastern and western parts of the city, large-housing estates had also been established since the 1960s, as a particular outcome of state planned urbanization and housing supply.

Historically, despite the high degree of change that characterized particularly the twentieth century, regional development in general and suburbanization in particular had followed a kind of time-space continuum that ties up to the de-concentration of land use and the conflicting interests between core city and periphery. Even in the beginning of the twentieth century, at the latest with the completion of the administrative reform and the creation of "Groß-Berlin" in 1920, the industrial city of Berlin achieved not only spatial expansion and high densities, yet also developed significant connections with the surrounding areas. The current pattern of suburbanization still depicts to these historical transformations, since it follows the different corridors heading to the north and northeast, the east and southeast, and the west and southwest according to the early development corridors of the "S-Bahn"-system that had been built about 100 years ago.

Industrial suburbanization in and around Berlin took off in the centre of the city and reproduced a periodical pattern of locational shifts for a time span of at least 60 to 70 years. The starting point of such kind of movements were firms of the heavy industries, such as machine tool manufacturing, who needed to leave behind the critical situation of the urban core (Zimm 1959). Zimm (ibid.) distinguished up to five different phases in which specific locational dynamics unfolded depending upon the availability of land and the improving traffic conditions (see Table 4.2). The latter applied first to the influence of the waterways, afterwards then to the railways.

Since the establishment of Groß-Berlin in 1920, suburbanization ocurred in the context of several "internal" shifts within the core area of Berlin (Bernhardt 1998; Zimm 1959). Such internal shifts have to do both with the decentralization of households and their movement into small suburban towns or even villages, and also with the suburban dynamics of the industry. These spatial shifts have inscribed a basic pattern of suburbanization into the structure of the region which is still visible today. According to Zimm (1959, 211), the pattern anticipated continuous and discontinuous developments that ocurred afterwards. In order to explain this pattern, the author suggested namely the role of land prices, transport situation and the concentration of population: each of these factors related to the industrial development. The increasing growth pressure of the industries to move out of Berlin have been touching also those subareas that had originally been developing as neighbouring towns of the coming Groß-Berlin. Industry then began to develop strong spatial linkages that support the view that these towns were developing to industrial satellites in the sense of Taylor (1915, see Chapter 1).

Later on, Zimm had put it as followed: "If looked at the locational setting of the surrounding districts in the Berlin area, the suburbs were developing the function of a *complementary resource space* (emphasis M.H.). This was mainly caused by the shift

Table 4.2   Localization and movement of industrial sites in Berlin

Phases	Major players	Locations	Determinants
1: 1800	City-state (Submission of user permits and commercial licences)	Urban core	Concentration of workforce
2: 1800–1840	Textile industry, machine tool industry	Edge of the core: Oranienburger Tor	Workforce, local iron foundries
3: 1850–1871	Machine tool industry	Chausseestraße, Disctrict of Moabit	Land prices, access to horse-tram
4: 1871–1895	Machine tool industry, elektrotechnical industry	S-Bahn periphery: Northern inner-city	Land prices, railway, housing
5: 1895–1918	General industries, Heavy industries	S-Bahn periphery: Districts of Wedding, Charlottenburg, Lichtenberg, Schlesisches Tor	Land rents, S-Bahn, waterways
	Heavy industries, General industries	Districts of Tegel, Lichtenberg, Tempelhof, Marienfelde, Ober-/ Niederspree	Land prices, S-Bahn-network, railways, waterways

Source: own after Zimm 1959, 212

of industrial sites, the development of agricultural land, the provision of housing for commuters to Berlin and also recreational functions that were provided for the entire region. However, these functions, emerging from a spatial division of labour between city and suburbs, had been significantly determinated by Berlin." (Zimm 1987, 138)

Keeping in mind several historical ruptures to the process of urbanization which makes the case of Berlin-Brandenburg quite unique, the more recent developments since the political turn in 1989/1990 are characterized by three factors: first, by economic transformation on both sides of the former border, in the course of de-industrialization and sectoral change. The high job losses in the industry are nowhere near being offset by the tertiary sector. Yet there are many indications that the service sector – for example, craft businesses, producer-oriented services, and bicycle couriers – will be a key factor for the future economy. The general trend towards structural change in logistics (demand rather than supply markets, general outsourcing of services, reduction of stocks and warehousing, logistic integration of various stages of added value) is emerging in a particular manner in this region.

Second, this applies to the suburbanization process that occupies a central position here. There are two levels at which trade and settlement structures affect logistics and freight transport and *vice versa*: on one hand, the exodus of commerce and manufacturing causes the diversion of freight transport directly supplying suburban companies and residential areas. On the other hand, the periphery takes over an increasingly substantial part of the commodity-handling function of the core city as the handling and distribution companies move away from the cities and into the surrounding areas. The changing location structure also affects the configuration of the logistics chains, especially with regard to the long-haul/short-haul transport distribution ratio.

Third, the economy of the Berlin-Brandenburg region still displays a below-average regional orientation – even many years after unification. The delivery networks of Berlin's enterprises are traditionally strongly oriented towards western Germany, the old hinterland in Hamburg, Lower Saxony and even North Rhine-Westphalia, thus artificially expanding many logistic chains. At the same time, this region is now a huge consumer area that produces disproportionately few goods and therefore has a transport imbalance. Both factors are only – if at all – likely to change in the wake of comprehensive economic renewal and the regional integration of added-value stages and chains.

Although development had already taken place along the backbone highway (A 10/110) and the Berlin beltway autobahn before, the process of change has speeded up from the 1990s. The new suburbs at the Berlin urban fringe consist of housing, retail, and commercial development (see Table 4.3). Freight transport, trucking and warehousing firms have been actively attracted to locate in newly developed dedicated freight centers, established at three strategic locations in the West, South, and East of the City of Berlin. The completion of the freight centres will alter the landscape of the physical distribution facilities around Berlin.

Table 4.3   Total occupation in Berlin and suburbs, 1993–2004

	1993	1998	2001	2004	1993–2004	
					absolute	in %
Berlin	1,337,366	1,132,570	1,125,714	1,035,943	-301,663	-
Suburbs	289,006	303,939	324,542	283,504	5,502	

Source: Arbeitsagentur Berlin-Brandenburg, own calculations

Locational Dynamics of Logistics and Freight Distribution

The Region as a Distribution Area

In the aftermath of the post-1990 political-economic transformation the region has undergone a spatial re-configuration between the core city of Berlin and in the newly emerging suburban areas. Regarding freight flows, the region-specific supply and demand structures have created a complex transport pattern. The pressure of structural change has led to changes in the origins and destination of freight flows and the division of responsibilities of freight forwarders (cf. IVU 2000). In 1992, the volume of goods transport in Berlin amounted to 89.9 million tons, dropping to 87.4 million tons in 1998; in Brandenburg the volume fell from 286.2 million tons in 1992 to 262.7 million tons in 1998. This initially surprising decline is due to de-industrialization in both states, the increasing importance of the tertiary sector and the steady drop in construction activity during the second half of the 1990s, which was the main reason for the decline in the bulk-cargo and basic-industry sectors. At the same time, the modal split was changing. Now road transport is being used for about two-thirds of the volume of transport and about four-fifths of the entire traffic in both subregions. Transport by rail and inland waterways is stagnating at best. In terms of spatial orientation the trend is balanced: The slump in industry in the 1990s, with a 46 per cent drop in jobs in manufacturing industry between 1991 and 1999, is still offsetting the rise in freight transport due to increased trade with other federal states.

Berlin's links with other economic regions continue to reflect the city's decade-long economic isolation from the areas surrounding it. It still imports considerably more goods than it exports (a ratio of 1.7 to 1, or 58 million tonnes of incoming as opposed to 34,7 million tonnes of outgoing goods). Berlin's most important trading partners – apart from the state of Brandenburg – are North Rhine-Westphalia, Lower Saxony, Saxony-Anhalt, and abroad (BMVBW 1999). Almost 75 per cent of domestic imports and more than 80per cent of exports in 1998 were transported by road and a substantially lower percentage by rail and inland waterways (ZLU 2001). The share of rail freight is much smaller here than on the pan-European scale, which accounts for 12 per cent of imports and 15 per cent of exports (ZLU 2001).

In the mid-1990s – after the first wave of post-1990 industrial relocation had ebbed – it emerged that the big industrial beacons (*Daimler-Benz* in Ludwigsfelde, *BMW-Rolls Royce* in Dahlewitz) would be the exceptions to the rule, which tended to consist of the construction trade, services, wholesale and retail trade. The freight trade also belongs to this mixture of light industries and development also occurred in waves after unification. In the early years, when institutional restructuring was proceeding only slowly, the freight forwarding branch made a good start and established large-scale developments, for example in Falkensee and Brieselang on the western motorway ring or in Genshagen, directly alongside the southern motorway ring. Once planning and development of integrated freight centres

(IFCs) had begun, treatment of logistics companies became more restrictive; with only a few exceptions, the state of Brandenburg stopped subsidizing logistics locations outside the IFCs. The obvious aim was to concentrate logistic companies in the IFCs and to recover initial public investments as soon as possible.

Attitudes became more open towards this branch of industry after de-industrialization had exacerbated the economic crisis in eastern Germany and investment activities had slackened around Berlin. The current economic situation has put pressure on planning policy to canvass for as many customers as possible. In 2005/06, after Brandenburg's state government had revised its funding philosophy, logistics was declared a key focus of its regional development policy. In cooperation with Berlin's governing Senate plans are being made to establish a Berlin-Brandenburg logistics network to complete marketing the IFCs and to strengthen the profile of logistics as a pillar of the region's economy. In logistical terms, the individual subareas of the study area can be described as follows (see Map 4.1):

- The inner-city is a central consumption area to which commodity flows from households, retail and services are delivered. Handling occurs only on the "last mile", that is, just prior to (and largely by) the recipient or end-consumer, previously also on the premises of Hamburger Bahnhof and Lehrter Bahnhof (HuL handling station), which has been relocated afterwards to the IFC at Großbeeren.
- The important areas of the urban periphery are the district centres, residential areas and undeveloped areas as well as Berlin's major belt of trade and industry, known as the Berlin commercial ring, that stretches from the north-west (Tegel, Reinickendorf) via the north-east (Pankow, Marzahn, Lichtenberg) to the south (Lichtenrade, Marienfelde). These production and distribution areas arose during Berlin's first two waves of inner suburbanization before and after the turn of the last century (SenStadt 2000).
- In the Brandenburg part of the Zone of Mutual Interdependence a large number of new handling locations have arisen since unification (at about 60 large locations from a total of some 110 industrial areas), partly owing to out-migration of Berlin enterprises and partly because of new companies from elsewhere. The activity range of this subarea is also twofold: some locations only deliver freight to the Berlin area – some are also involved in business unrelated to Berlin.

The logistics companies mainly consist of freight forwarders, haulage firms and warehousing companies. The businesses represented in the new industrial areas include a few wholesalers, of which some have relocated from Berlin and some have moved here from outside the region or have built up a completely new distribution structure. Among these are the distribution centres of the retail chains; some of these moved to the IFCs, some were built at new sites elsewhere. The

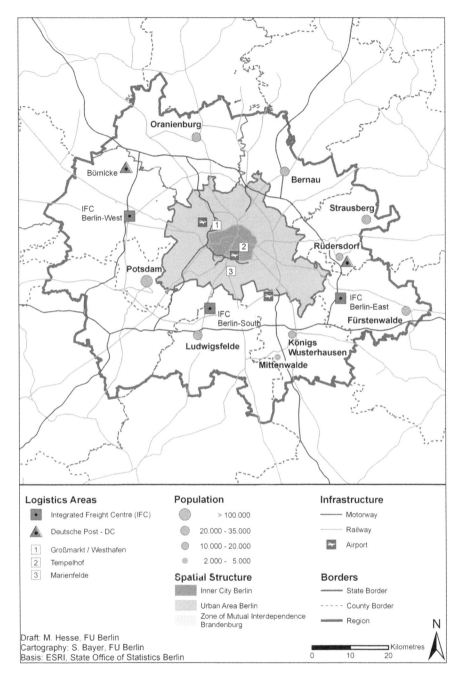

Logistics Areas

■ Integrated Freight Centre (IFC)

▲ Deutsche Post - DC

1 Großmarkt / Westhafen
2 Tempelhof
3 Marienfelde

Population

● > 100.000
● 20.000 - 35.000
● 10.000 - 20.000
● 2.000 - 5.000

Spatial Structure

▨ Inner City Berlin
▨ Urban Area Berlin
▨ Zone of Mutual Interdependence
   Brandenburg

Infrastructure

—— Motorway
—— Railway
✈ Airport

Borders

—— State Border
- - - - County Border
━━ Region

Draft: M. Hesse, FU Berlin
Cartography: S. Bayer, FU Berlin
Basis: ESRI, State Office of Statistics Berlin

Kilometres
0   10   20

N

Map 4.1   The Berlin-Brandenburg Region

latter include Altlandsberg on the north-east Berlin Ring and the Orion industrial estate in Kremmen in the north-east, one of the few logistics sites outside the IFCs that received federal state funding for reasons of structural policy.

Some 40 kilometres away from central Berlin is the town of Mittenwalde. There a food chain has established its distribution centre for all eastern Germany (including Berlin) and other distribution companies have settled since then. Whereas nowadays freight forwarders still count as partially attached to the core city, inner-city distribution centres for the retail trade tend to be the exception rather than the rule. At the present time there are only two known cases of new distribution centres in urban locations: in 2000 a retail company built a distribution centre on 16 hectares of land in the Pankow-Nord industrial area of Berlin to serve Berlin and the surrounding area. In 2004 a local firm built a new distribution centre on a former gasworks site in Marienfelde (south-west Berlin) to replace three different locations in the same district.

The Development of Logistics Employment

The spatial distribution of the distributive services in the Berlin-Brandenburg region is described on the basis of the number of employees subject to social insurance contributions according to the official German classification of branches of economic activity (WZ93, WZ03) for the 1998–2005 time period. The term *logistics* comprises those sectors that, in statistical terms, are clearly separated from passenger transportation and also locationally flexible, i.e. their location decisions do not depend on waterways or railways. They include freight transport by road (section 60.24), cargo handling and storage (63.1) and freight forwarding (63.40.1). In addition, the corresponding data on the development of the wholesale trade (51) were included. Wholesale trade is an important part of physical distribution and – in terms of economic systematics – also includes the wholesaling activity of retail chains. Data from the previous classification (WS73) were incorporated yielding a time series dating back to 1993, although the two periods are not exactly comparable owing to differences in classification criteria. 1993 was the first year with full statistical coverage of developments in the Neue Länder after German unification.

Regional employment in the logistics sector has been declining in the entire Berlin-Brandenburg region since 1998. This trend is even clearer when 1993 and 1995 are included despite the different classification criteria in these two years (see Table 4.4). Whereas the sectors under consideration had grown considerably throughout Germany up to 2001 and have stagnated since then, the Berlin–Brandenburg region has not appreciably benefited from a growth in logistic services. This may well have to do with the post-1990 loss of almost 200,000 jobs in industry in Berlin alone and also with the region's hitherto ineffective – but much-evoked – function as a hub for trade with the countries of central eastern Europe.

Table 4.4   Employment in selected groups of logistics in Berlin and suburbs 1993–2005, also in Germany

	1993	1995	1998	2001	2004	2005	98–05
Berlin	20,836	19,716	14,575	13,619	12,046	11,812	-17,4
Suburbs	6,243	8,420	10,646	14,386	12,940	12,921	21,4
Total	27,079	28,136	25,221	28,005	24,986	24,733	-1,9
Germany	564,943	580,525	596,395	661,366	656,623	652,217	9,4

Source: Bundesagentur für Arbeit, own calculations

1993/1995: WS73 (651: Road freight transport; 670: Freight forwarding/warehousing)
1998–2005: WZ93/WZ03 (6024: Road freight transport; 631: Warehousing/storage; 63401: Freight forwarding)

There has been a significant shift of employment from the core city to its surrounding area: in 1993 Berlin still had more than 20,000 people working in the logistics sector, plus more than 37,000 employees in wholesaling. By 2005, the number of jobs in logistics had dropped to fewer than 12,000 – a significant decrease of almost a third compared to 1998. In contrast, the Brandenburg part of the Sphere of Mutual Interdependence was able to increase its share by more than one-fifth between 1998 and 2005, and more than double after 1993. However, this was not a steady increase: it peaked in Berlin's surrounding area in 2001, but then the general downturn also affected jobs in the logistics branch. Since 2001, however, both subareas have been more or less equal (measured in absolute figures) in logistics employment. In 2005, for the first time, Berlin's surrounding area achieved higher figures in absolute terms than the core city. The rising importance of the surrounding areas is being confirmed by the location quotient. The location quotient reveals the regional significance of the analysed sector, since the regional performance of logistics employment is set in relation to the development of that particular sector in a reference area, in this respect Germany as a whole. Thus the specific strength or weakness of the regional can be identified. It is evident that the Berlin area lost significance and the suburbs increased their share correspondingly mainly during the second half of the 1990s (see Table 4.5).

Not all subareas profited to the same extent from the surrounding area's increasing importance (see Map 4.1). The biggest boosts in commercial and industrial development occurred along the southern Berlin ring. This mainly applies to Ludwigsfelde (Preußen-Park for instance, an old industrial site that has been successfully regenerated), Brandenburg-Park in Genshagen and IFC Berlin-Süd at Großbeeren. A second, smaller focus is the Fredersdorf/Vogelsdorf area in the north-east, where both sides of the Berlin ring accommodate industrial areas with a high percentage of trade (wholesale and retail), transport and logistiscs (including *Deutsche Post*'s freight centre for eastern Berlin and south-eastern Brandenburg).

Table 4.5   Location quotient of logistics employment in Berlin-Brandenburg

	1993	1995	1998	2000	2004	2005
Berlin	0.8	0.762	0.591	0.522	0.461	0.463
Brandenburg	1.1	1.333	1.591	1.783	1.865	1.855

Source: Bundesagentur für Arbeit, own calculations

1993/1995: WZ73; 1998–2005: WZ93/WZ03

The western axis along the B 5 also counts as a development corridor along with the Berlin-West IFC in Wustermark, another big logistics centre near the Berlin ring-road (Brieselang), and *Deutsche Post*'s freight centre for western Berlin and western Brandenburg. In comparison with Berlin's industrial development as a whole, the northern part is lagging behind; apparently (West) Berlin's traditional orientation towards western Germany has been a disadvantage here. Industrial and commercial locations are restricted to both sides of the motorway ring.

Overall, the distribution sector accounts for a considerable share of suburban industrial development. This impression gained from employment statistics is confirmed by a glance at the distribution of the individual branches located in the suburban industrial areas. Regarding the development of the individual industrial locations in the area surrounding Berlin, the distribution/logistics branch occurs most frequently in the locations considered with almost 21 per cent of all mentions (by companies) followed by manufacturing and producer-oriented services (Aengevelt Research 1999). The locational strategies of such firms immediately reflect the particular land rent differential between city and suburbs (see Table 4.6). If wholesaling is included in the distributive services, almost one-third of all enterprises in suburban industrial areas belong to this sector. If we also consider that this branch belongs per se to the most space-extensive sectors, the importance of logistics and distribution becomes very clear.

Table 4.6   Rents and land prices for office-, service- and warehousing space in Berlin-Brandenburg (in euros per square metre and month; in euros per square metre)

	Office rents	Services	Warehousing	Land prices
City of Berlin	11	6–7.5	2.5–6	
Suburbs	6–8	6–7.5	2.5–4.5	25–100
Business Parks	6.5–8	7–8	3.5–5	

Source: Jones Lang LaSalle 2004/2007

Selected Spaces of Distribution

The locational shift from core Berlin to the surrounding area, along the ring motorway, or to one of the new suburban distribution centres is occurring against the backdrop of specific demands by businesses and the respective appropriateness of core-city and suburban locations. The companies' decision-making processes and calculations vary greatly in this respect, depending on the type of distribution, on company integration into overarching logistical networks and also on the availability of certain locations at a given time. As an intermediary analytical step between assessing the spatial distribution of the employees (macro-level) and reconstructing the decision-making process of the individual company (micro-level), an outline will be given of the history of three typical locations, each of which has a high percentage of logistical uses: a core city area in Berlin, the three freight centres in the area around Berlin and the Ludwigsfelde complex at the southern ring motorway. This approach will lend transparency to the "spatial logic" within which each location developed at its respective site.

Logistics Area at Westhafen/Großmarkt/Saatwinkler Damm (Berlin)

The logistics area at Westhafen/Großmarkt/Saatwinkler Damm consists of three separate locations and is situated at the northwestern edge of Berlin's inner-city. In the industrial development plan of the Senate Department for Urban Development, the entire complex is designated as a freight transport corridor and sub-centre supplying Berlin's inner-city. The complex comprises Westhafen (west port), the Großmarkt (wholesale market) in Beusselstraße and the industrial areas at Saatwinkler Damm and Friedrich-Olbricht-Damm in Charlottenburg. This area has excellent rail, waterway and city motorway links and directly adjoins the city centre. Altogether with the former neighbouring container terminal at the Hamburger und Lehrter Bahnhof (HuL) and the freight forwarders located in nearby Heidestraße, this area encompasses the logistic centre of what used to be West Berlin. In the postwar years, logistics was of paramount political importance to guarantee the supply of food and resources to the then isolated city. Now all three locations are under strong pressure to change: especially Westhafen and Großmarkt; Saatwinkler Damm/Friedrich-Olbricht-Damm are among the key urban industrial areas in the Berlin Senate's urban development plan for commerce and industry.

The Westhafen port facility went into operation at the northern edge of the city centre in 1927. It can easily be reached by the A 100 motorway and at the same time has good connections to the inner city. Westhafen is a typical inner-city port that is undergoing adaptation processes in inland water transport and the handling sector (shift to larger transport units, quicker handling cycles, distribution instead of warehousing) on the one hand, and is exposed to growing competition and compatibility conflicts on the other. Unlike Osthafen (east port) in Friedrichshain, Westhafen has been designated for longterm logistical use. The commodities handled here include solid and liquid fuels, construction materials and other bulk goods. As a part of an extensive investment programme during the past few years, conditions have been created in Westhafen to enable logistical modernization and hence to improve competitiveness. In the long term, it is planned to make the port into a trimodal logistic centre for handling inland waterway, rail, and road transports. The third dock was filled in to create space for new buildings (especially warehouses and handling facilities) for freight forwarders. The port has its own port railway and ro-ro facilities. In addition, a container terminal started operation in 2002. Port infrastructure includes 93,000 m^2 of covered space a total of 173,000 sqm of outdoor space, grain and cement depots, and a heavy-duty crane with a load capacity of 350 t.

After completion of the Magdeburg waterway crossroads and the new Charlottenburg Lock as part of the German Unity Transport Project No. 17 (extension of the Elbe, Weser and Rhine shipping route), Westhafen will be accessible in the near future for Europe ships with 2,50 m draught and large-engine cargo vessels with 2,20 m draught. After further expansion large-engine cargo vessels with a draught of 1,50 m should be able to enter Westhafen from 2008 onwards (IHK zu Berlin 2003). Berlin's Port and Warehouse Company (*Behala*) had also suggested relocating the terminal for combined rail traffic from Hamburger und Lehrter Güterbahnhof (freight station) to Westhafen to enhance the latter's interface status. Meanwhile, however, the *Deutsche Bahn*'s freight transport subsidiary has moved the HuL transports to the IFC at Großbeeren. The upgrading of Westhafen to a logistics centre has met with public criticism; in particular, misgivings have been voiced that more warehouses would generate an increase in truck traffic (which reaches substantial levels at all port locations) and hence lead to higher stress levels for local residents.

The Westhafen cluster also includes Berlin's Großmarkt (wholesale market) at Beusselstraße in Moabit. The market, about 33 hectares in size, replaced two sites in Charlottenburg and Mariendorf (West Berlin). Its predecessor was the first wholesale market hall at Alexanderplatz, which opened in 1886 and soon reached its limits. In 1913 Berlin's governing body had purchased the land of what is now the Großmarkt at Beusselstraße. However, plans for a central food market were foiled by two world wars and the subsequent partition of the city. Planning did not start again until 1960, and the market opened in 1965. Since the Wall came down in November 1989 Berlin's Großmarkt has expanded and is now also supplying the surrounding regions of Brandenburg. According to the operating company, *Berliner Großmarkt GmbH*, proposals to relocate the wholesale market to outside Berlin have been abandoned, because its present central location and easy accessibility offer big advantages over a peripheral site.

The sales area of the Großmarkt wholesale market covers almost 50,000 m^2 and is used by about 140 companies. Some 2,500 employees work in the precinct. Almost one-third of the total area is taken up by buildings, including sales and storage areas, meat packing plants, cold-storage and refrigeration rooms, banana ripening chambers and office space. As well as the wholesaling companies, the market's tenants include freight forwarders, repair firms, public services offices, petrol stations, vehicle wash facilities and recycling firms. The location has the special advantage of being near to the eastern and western city centres and easy to reach via city motorway. It has its own railway siding, but this had not been used for years. According to the owner, the *Fruchthof Berlin Verwaltungsgenossenschaft*, 99 per cent of goods are delivered and collected by truck; there are no plans to use the railway again. Now that the major food chains have modernized their logistics and built their own distribution centres, however, the wholesale market has become less important:

> All Berlin's major food companies used to concentrate their fruit trade on Berlin's Großmarkt. That was where fruit was delivered, handled, then consigned to the halls of the big suppliers and trucked off again. That's all over now. None of the big suppliers handles fruit at the wholesale market any more. Only the wholesalers are still there. Some of the wholesalers have left Berlin too. So basically, if you look at Berlin's wholesale market, all you see are a lot of Turkish tradesmen selling their fruit. [...] There's only a few left. *MeyerBeck* used to be at the wholesale market, *Rewe*, too; all of them were there and now there's nobody, we were the last to leave. The chains used to be the big buyers, now there's only *Hase* and *Hagebau*, *Metro*, and *Aldi* a little bit, they're the last big company. Wholesalers used to be big there, they've all gone too. (LM1: logistics manager)

There have been public complaints about Westhafen and Großmarkt for some time now, because their access road, Beusselstraße, runs right through the Moabit district and is one of Berlin's busiest streets with a share of truck traffic, so operation of the two sites involves high noise and pollution levels. This demonstrates the fundamental conflict between the potential inherent in the traditional situation and infrastructure of these locations and the conflicts generated by modern logistics and freight transport – mainly due to the growing importance of truck haulage in the logistics chain and to the pressure for locations to expand. For this reason, measures have already been started to lessen the impact of traffic on Beusselstraße, especially a night-time ban and a 30 km/h speed limit for goods vehicles. It is to be expected, however, that the Senate Department for Urban Development's plans to upgrade Westhafen to a subcentre for freight transport will be closely curbed by the low tolerance levels in the neighbourhood.

In the northwest, the industrial location at Saatwinkler Damm/Friedrich-Olbricht-Damm in Charlottenburg has not such problems, being surrounded by the city motorway; its access roads do not adjoin residential areas, and the industrial sites are bordered on both sides by allotments. The Friedrich-Olbricht-Damm is a traditional industrial location with excellent links via the city motorway, which is why transport businesses in particular prefer it. Currently its tenants are including wholesalers and retailers, as well as several transport firms (forwarders, parcel services, freight transport). The Saatwinkler Damm is a fairly new industrial area near Friedrich-Olbricht-Damm and is still being developed. The site occupies about 13.7 hectares and was used to store coal from 1955 to 1989. In future, it is to be used as an industry and commerce quarter, especially by small and medium-sized manufacturing and service companies. Berlin's Senate has highlighted the importance of the quarter by including it in Berlin's Industry and Commerce Urban Development Plan (BIG 2000), so it can be considered to be a prototype for future policy in this field:

> The creation of modern services structures presupposes a stable industrial basis. Only the intelligent combination of industrial products and services can produce competitive goods. Many production- and industry-related services are only able to develop their full potential when they are separated from the manufacturing enterprises and provided by specialist firms. As the economy continues to specialize large-scale industrial complexes are being replaced by many small and closely interconnected firms that supplement each other. At the same time the boundaries are becoming fluid between conventional production and the service sector. A growing number of companies are becoming urban-compatible in terms of both areal requirements and environmental criteria. Against this backdrop one of the priorities of regional economic policy must be to create the necessary infrastructure for developing small-scale networks of research institutions, manufacturing firms and industry-related service providers. This applies especially to Berlin where – owing to post-WWII constraints – the enterprises are much less interconnected than in the other economic areas. (Senate Administration for Economy and Labor 2001)

Hence, all three locations of this cluster may be seen as representative components of an integrated concept intended to keep manufacturing industry and related service providers in Berlin. The distributive sector also belongs to this category, the branches "freight transport by road" (60.24), "cargo handling and storage" (63.1) and "freight forwarding" (63.40.1) being explicitly mentioned in the concept (Senatsverwaltung für Wirtschaft und Arbeit 2001). However, the Senate departments are also aware that this specific sector contributes to considerable problems and adversely affects the goal of urban industrial policy.

Integrated Freight Centres (IFC) Berlin-East, -South, -West

After unification the distribution networks for consumer goods were restructured in the Berlin-Brandenburg region. In the process, regional distribution centres have developed either as logistic warehouses or as freight forwarding facilities that organize a large part of the region's goods supply (furniture, food, mail-order and department-store goods, new vehicles, etc.). Public planners were also among those involved in this process: In the Brandenburg part of the Zone of Mutual Interdependence three large integrated freight centres (IFC) were built in relative proximity to Berlin: in the municipalities of Wustermark (West), Großbeeren (South) and Freienbrink (East). They are intended to contribute to shifting transports from road to rail and thus reducing the adverse environmental impact of the freight traffic heading for Berlin. In addition, they are built to serve as logistic hubs in organizing the city's logistics distribution, that is to optimize Berlin-related deliveries.

The IFC Berlin-West is located 16 km west of Berlin's inner city in the municipality of Wustermark. It links road, rail and waterway owing to its position at the intersection of the Berlin outer railway ring and the main railway line between Berlin and Hannover, the adjacent Havel Canal, Berlin's western motorway ring (BAB 10) and the B 5 (see Map 4.2). Plans envisage a gross total area of 265 hectares and a net floor space of about 160 hectares, which will accommodate the distribution centres (forwarders, logistics, trade), a container terminal and (in the long term) a port. Additional service facilities such as central IT, gastronomy/ hotels, a customs office, a post office and a truck stop area have not yet been built. The four-lane development of the B5 federal highway guarantees road access. By September 2006, 24 companies which employ a workforce of 1,564 people had moved to Wustermark, mainly from the freight forwarding and logistics sector and trade distribution (data according to the developer).

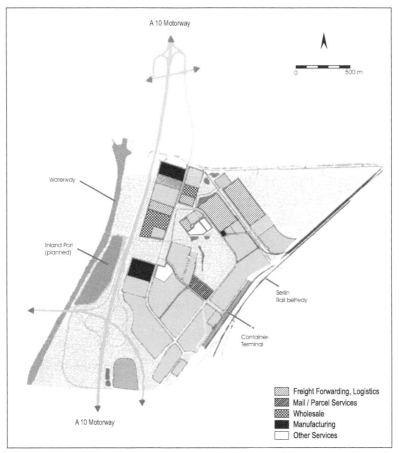

Map 4.2   Land use and infrastructure of the IFC Berlin-West

Source: Own

The IFC Berlin-South is situated 22 km south of central Berlin. It is connected to the German motorway network via the Berlin ring motorway and to the Berlin road network via the B101 federal highway. The centre has a main track rail connection and a link to the outer Berlin rail ring. The IFC has a gross area of 260 hectares, of which about 150 hectares floor space can be leased. Its facilities include a *Deutsch Bahn* container terminal, a logistics park for SMEs, a container service centre and service facilities (see Map 4.3). The IFC railway station has been used by *Deutsche Bahn* since 2003 in the wake of the HuL terminal's relocation to Großbeeren. However, forwarders and shippers make only limited use of the facility because of its high prices, poor transport connections and loading times. By September 2006, 41 companies had been investing in properties in the IFC, which also host 13 tenants in addition, with a total of 3,180 employees (data according to the developer).

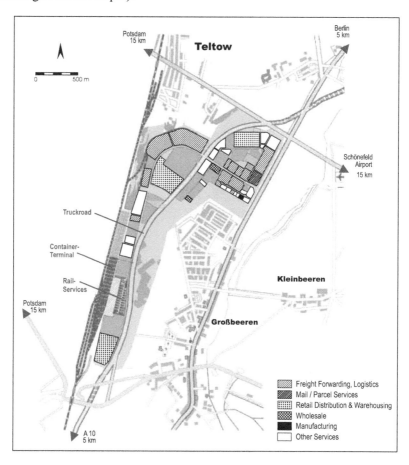

Map 4.3   Land use and infrastructure of the IFC Berlin-South

Source: Own

The IFC Berlin-East (Freienbrink) is located in the southeast of Berlin, 5 km away from the city limits. The site used to be the material supply depot of the Ministry of State Security in the former GDR. Plans for the centre envisaged interlinking of wholesale distribution, freight forwarding and an industrial park. Its area totals 130 hectares with 62 hectares available for commercial logistics, another 27 hectares for manufacturing industry and 5 hectares for freight forwarders. It can be accessed directly from the Berlin motorway ring (BAB 10) and via the L 38n. The IFC has its own feeder line and offers future tenants ten parcels of land with their own sidings; facilities such as a service centre and a truck stop area had originally been planned, but have not been realized. By September 2006, 20 distribution companies were located in Freienbrink accounting for a workforce of about 1,170 employees (data according to the developer).

These three locations are typical large-scale metropolitan-area integrated freight centres with considerable space requirements; their primary functions are to provide the transport industry with space at a well-developed and easily accessible location with minimal disturbances. All three locations are designated commercial areas (*Gewerbegebiete*, GE) or industrial areas (*Industriegebiete*, GI). The only conditions for investors and users relate to the minimum and maximum sizes of land for acquisition; the originally envisaged contracts to regulate logistic organization (use of intermodal transport, participation in collaborative activities) have not been put into practice.

Because of the high space requirements and the concomitant long distances between the IFC locations and the delivery addresses, there are reservations about the concept of large IFCs being the optimal solution to improving urban distribution. Their peripheral location generally means long transport distances to and from the site, which is often a crucial counter-argument, apart from the locational decisions already made by the individual forwarders. In addition, these long distances also exacerbate traffic and thus the environmental performance of the IFCs. A possible solution has been suggested: logistics sub-centres (LSCs) within the city; these interfaces would provide a high quality of logistics service in immediate proximity to the freight volume. LSCs represent short distances before and after handling and distribution; they exploit the advantages of intermodal transport and modern metropolitan-area logistics without the pollution and locational disadvantages associated with IFCs. In addition, they would enable a compact mix of transport, logistics and service companies. In early 1996 the first LSC was established on 7 hectares of land at Neukölln/Treptow freight station and went into operation in summer 1997. Its business includes rail deliveries for a large department store and storing green coffee beans for a coffee roasting facility.

Estimates based on freight transport statistics indicate that some 20 per cent of the road freight traffic entering Berlin passes through the IFCs, the remaining four-fifths are handled by transport and logistics companies that are located outside the IFCs – both in Berlin and in Brandenburg state (the sphere of mutual influence and the rest of the state). The latter volume is estimated at about five million tonnes per year, whereas the total volume of goods transported to Berlin by road was about 38 million tonnes in 1998 (ZLU 2001).

With these three locations, the Berlin-Brandenburg metropolitan area had evolved into one of the most advanced locations of IFC start-ups in Germany by the late 1980s and early 1990s. Initial problems finding occupants at the three Berlin locations have been largely overcome. By September 2006, more than 98 companies (tenants included) were located here, and some 6,414 jobs had been newly created or transferred (see Table 4.7). In the coming years it is anticipated that the now-vacant parcels of land will be occupied and the remaining gaps in the road links will have been closed.

It is not clear yet which planning effects may be triggered by the IFCs. In the implementation process under market conditions, considerable contradictions and conflicting aims have emerged between the two macro-goals of location development: transport optimization and provision of space. Building terminals and interfaces takes time and depends on market acceptance by potential occupants – and at exactly the right time. The pressure of competition means that the operators and/or marketing agents need to find buyers or tenants for their spaces in order to be cost-effective. The constraints of this logic mean that, first, there are fewer qualms about accepting companies that are basically unsuitable for such peripheral locations (such as craft businesses) and, second, it is unlikely that obligations will be imposed on the companies to use rail transport or to participate in cooperation initiatives to reduce traffic. Initially, the container terminal at the IFC Berlin-South had not been used for years. However, after the former central-city hub for combined transport had been transferred from HuL (close to the new Berlin main station) to the IFC Berlin-South, and since the container terminal

Table 4.7   Integrated Freight Centres in Berlin-Brandenburg

Location	Size (gross)	No. of firms	Employees	Subsidies	Services	Infrastructure
Berlin-West	137 ha (226 ha)	24	1,564	96.0 m Euros infrastructure, 40.0 m Euros Firm subsidies	Freight forwarding, logistics, warehousing	Motorway, Container-Terminal, (Inland port)
Berlin-South	150 ha (260 ha)	55	3,680	94.7 m Euros infrastructure, 9.6 m Euros Firm subsidies	Freight forwarding, Parcel services, Grocery distribution	Motorway, Rail access, Container-Terminal
Berlin-East	96 ha (130 ha)	20	1,170	11.5 m Euros Infrastructure	Retail logistics	Motorway, Rail access

Source: own investigation based on data provided by IPG Potsdam/LEG Brandenburg; data of September 2006

in the IFC Berlin-West is operating as well, the number of combined shipments going into and out of the IFC is increasing significantly.

In general, the IFC trend appears to be a high-risk policy with few options for reliable planning targets. Supply planning of IFCs has shown that it is not necessarily possible to exert influence on transport organization, this being in the hands of the customers (the companies). Furthermore, the company itself experiences conflicts between the targets of short-distance and long-distance transport, making it difficult to assess the effects precisely, but also pointing to adverse impacts. As a rule, the option is to reduce long-distance transport at the cost of an increase in local traffic, which generally results in a negative total balance for the locations in question. A study of the traffic effects of logistics nodes (Sonntag et al. 1998) established the concrete impact of the IFC Berlin-South on the basis of simulations. Out of the six simulated variants only one (an IFC as mega-hub) achieved a positive traffic balance; in the other cases, the total volume of traffic lay above the so-called zero case (no IFC; company locations in the city area). In this respect, the following critical conclusion was drawn: "Logistics nodes primarily serve to optimize transport chains and to cut company costs (...). Logistics nodes do not a priori involve any urban/regional targets in respect of traffic and environment on the part of the users/operators. Such targets – if desired – have to be called for by the municipalities." (ibid., p. 136).

The problem the IFCs have with the concentration of heavy traffic is an immanent consequence of their generic property: to concentrate firms of the respective business at one locale. It becomes evident at the Berlin-South location, where the B 101 road between the IFC and Berlin has been partially widened and modernized but cannot, because of political reasons, be similarly developed on Berlin's side of the city limits. This will bring permanent problems for the companies because the bottleneck is steadily moving further towards the inner city and the tailbacks (into Berlin in the mornings, out of Berlin in the afternoons) cause delivery delays, especially at rush hour. The more the Berlin-South IFC fills up, the greater is the likelihood of increasing congestion on the B 101 into and out of Berlin.

Industry and Logistics Cluster Ludwigsfelde/Genshagen

The Ludwigsfelde location is the industrial centre of Berlin's surrounding area. Since unification in 1990 it has experienced a very positive development owing to relocations and new industries and also in the context of the recent historical development path. Although the population of Ludwigsfelde has remained static at about 22,000–23,000 inhabitants for some time now, employment has made an above-average upturn since the mid-1990s with more than 10,000 employees paying social insurance contributions. Together with Dahlewitz (also located on the southern Berlin motorway ring, but further eastward), this area had already been designated one of the few new industrial beacons of Berlin's urban area (Herfert 2003).

One of the key factors for this success was the restructuring of Ludwigsfelde's former major industrial location (an aircraft engine works before World War II, the *IFA* truck plant in GDR times) by *Daimler-Benz AG*, which currently produces transporters and vans here. After unification the former locations were regenerated, creating Industrial Park East and Industrial Park West with a total area of about 250 hectares. Another factor was the creation of two big commercial parks: the 220-hectare Brandenburg Park to the north of the motorway and the 110–hectare Preußenpark south of the motorway (cf. Herfert 2003). Before, another large industrial park was planned at the Genshagen motorway junction on the BAB 10 and two major Berlin freight forwarders, a wholesaler and a margarine company moved there in the early 1990s already. The rebuilding of an old site at Genshagener Straße produced another large location, supported in the early 1990s by investment by the Berlin branch of a freight forwarder with headquarters in western Germany. According to Herfert (ibid.) between 500 and 600 hectares of commercial and industrial space has been created in the town of Ludwigsfelde alone (including Genshagen, which was incorporated into Ludwigsfelde in 1993). This total does not include the Großbeeren freight centre (see above) situated only a few kilometres to the north.

Against the background of the historical development path, Ludwigsfelde has the advantages of industrial concentration and a favourable logistics location. Its three major plants (*Daimler-Chrysler*, *MTU* and *Thyssen-Umformtechnik*) have some 2,600 employees. In addition, there are many small and medium-sized firms, some of which supply the big three; others are completely independent. The two industrial parks and the other two industrial areas mentioned above are strongly service-oriented; the large firms include wholesalers, transport and freight forwarding services. Berlin is within easy reach via the southern motorway ring and the now completed B 101 road. There are also excellent links to long-distance transport.

Brandenburg Park is one of the most interesting of such concepts in Berlin's surrounding area. The industrial park approach was put into practice relatively exactly here. The concept is based on experience in the Anglo-American countries, especially England and North America, partly also France. Its developers intended to offer a very special product among the commercial and industrial locations in the sense of sustainable value and optimal presentation prospects. A key role is played by the park-like design of the commercial and industrial locations:

> All the firms that buy land here have to sign a ten-year maintenance contract for their land. As park developers we lay out most of the green zones and landscape zones on privately owned land, [...]. And in Germany the problem is that, if you analyse it exactly, private property owners plant their lawns wherever they want to, unless the zoning plan and building codes regulate it, but that's not usually the case. Maybe partly at the front, maybe partly at the front, at the back ...anywhere. With the zoning plan and building codes and especially with the purchase deeds we not only drafted this concept, this integrated concept, we kept to it too. (Brandenburg Park developer, 2002)

The companies pay a high price for this quality, sometimes much higher than comparable prices at other locations (see above). This is remarkable, especially because the percentage of freight forwarding and logistics companies at this location is not exactly small: and these branches are generally said to be low in capital and in danger of being squeezed out of the property market. The rapid marketing process in Brandenburg Park (in the early years at least) shows that there is a relevant demand within the quality market, even in this segment. This makes it clear that location decisions are definitely not ruled by cost, even in price-sensitive sectors, but that other factors also play an important part: especially the availability of space at the right time, possibly also quality-related criteria. The developer estimated that the run would have been better in the late 1990s if Brandenburg's policy had not enabled the neighbouring freight centre at Großbeeren to offer land at well below market prices, thus distorting competition.

Microperspectives on Corporate Decision Manking (n=50)

Methodological remarks In the two case studies company representatives and experts in the study areas were interviewed face-to-face. These talks were intended to supplement quantitative data and to elucidate spatial development dynamics in the context of entrepreneurial decisions. The idea behind the interviews was the analysis how companies act as a key determining factor in the context of commercial and industrial suburbanization. Therefore this research should inevitably focus on companies (cf. Markusen 1999). The second key factor in the process of finding and implementing a location is spatial planning, especially in conjunction with municipal and regional planning. For this reason, public planners, developers and estate agents were asked to give their opinions about the current development of Berlin's surrounding area and future prospects.

In both case studies, 131 people were interviewed. Eighty-three of these talks were based on an interview guideline for experts or on a semi-structured questionnaire for companies. On average these interviews lasted about an hour; the shorter ones were about 40 minutes long, the longer ones up to one and a half hours. The first interview series in Berlin-Brandenburg was conducted in autumn 2000 and comprised interviews with 24 regional experts and one company. In summer/autumn 2002 the second interview series was prepared and conducted with companies in Berlin-Brandenburg. It comprised guideline-based interviews with 50 companies for 53 locations; in addition, eight experts were interviewed. As a general rule, the interview partners were asked by phone if they were willing to participate; if so, they were sent detailed information (project outline, interview guide, questionnaire). Refusals came from eight of the 58 companies in Berlin-Brandenburg that were asked to participate. None of the planning and real-estate experts (see below) refused to take part.

When choosing appropriate companies and potential interview partners, care was taken to ensure that they were distributed almost evenly across the different sublocations in each region. On the one hand, multipliers such as the freight forwarders association of Berlin-Brandenburg or the Berlin chamber of industry and commerce were requested to suggest approachable entrepreneurs or executives. This procedure proved successful resulting in a participation rate of about 86 per cent of the companies asked. These also included firms with a reputation for persistently refusing interviews. Questionnnaires and responses from 50 firms for 53 locations and approximately 8,800 employees were evaluated. Of these, 14 companies or plants were located in Berlin (with about 1,200 employees), 39 in the Brandenburg part of the sphere of mutual influence (nine in IFC Großbeeren and six each in Ludwigsfelde and IFC Wustermark). The majority of the companies (28) belong to the freight forwarding and logistics branch followed by ten wholesale firms and seven food chains with their own distribution location. Six businesses belong to the CEP sector (courier-express-parcel services) (see Figure 4.1).

The semi-structured interviews were generally recorded on minidisc and later transcribed. The standardized results were entered in an evaluation matrix in Excel format, classed according to question category, and evaluated aggregately. The qualitative contents were categorized and evaluated using heuristic methods. In contrast to the California case study (see Chapter 5), companies in Berlin-Brandenburg were very willing to respond: only three of the interview partners refused permission to record their answers. In these cases the responses were entered into the semi-structured questionnaire.

In methodological terms, the qualitative interviews are closely connected with quantitative-statistic evaluations (see Flick 1991 and Lamnek 1995 on the role of qualitative surveys, and Friedrichs 1990 on empirical social research). They have a double function resulting from comparatively unstructured interviewing: First, qualitative interviews are intended to amplify and particularize the trends and structures that emerge from quantitative statistics, on the one hand, by scaled questioning, on

SPATIAL ORIENTATION

LOCATION	Core City	Suburbs	C+S	Region	Large scale
Core City	x		x x x x x	x x	x
Edge of Core	x x		x x	x	
Suburbs	x	x	x x x x x x x x x x	x x x x x x x x x x x x	x x x x x x x x x x
Suburbs beyond Motorway ring			x	x x x x	

Figure 4.1   Cases of firms studied in Berlin-Brandenburg

Source: Own

the other, by reconstructing the companies' decision paths and motivation. Second, to complement the quantitative methods, it is intended to obtain information that only emerges through the specific discussion of the subjective assessments of planning authorities and experts. If only structured and standardized methods are used, however, there is the danger that relevant areas will be left out, also if the corresponding response categories are not explicitly included or if adequate pre-tests are lacking (Flick 1991). Such a method has to be ruled out here because of the interviewees involved and the research approach employed. In this respect, qualitative interviews can bring added value if they are designed to amplify the interrelations that emerged in the standardized procedure. In this sense, qualitative interviews also serve to further differentiate the quantitative outcome; also, their purpose is to ask questions about categories that cannot be reconstructed via structured questionnaires because they are based on subjective knowledge and the interviewees'' own appraisals (regarding the assessment procedure, see Figure 4.2).

Corporate Decision Making – Contrasting Ideal Types

Freight Forwarder 1 (SPED1): Berlin Inner-City, District of Tempelhof This company is the Berlin branch of one of Germany's major medium-sized freight forwarders, founded in 1930. The Berlin location has existed at different addresses since 1951; the company has always preferred inner city sites. With its large industrial areas in a central location, Tempelhof can supply the entire city. The company currently has several locations: food logistics in Tempelhof, warehouses

1. Theoretical assumptions	
2. Pre-hypothesis	3. Breakdown of the subject into categories => key categories
4. Establishing a sample design	
5. Reconstruction of site selection behaviour through interviews	
6. Single case analysis	[Check of plausibility against the statistical background]
7. Case specific comparison and assignment (similarities, differences, meta-structure)	8. Interpretation of the cases alongside empirical substance
9. Comprehension of subjects to types that are more analogous regarding specific properties than others are (area of characteristics)	

Figure 4.2   Procedure of qualitative investigation and heuristic assessment of the findings

also in Tempelhof, and a freight transport base for contract logistics in Neukölln (the old West Berlin location).

SPED1's main Tempelhof location, with 16,000 m² and some 170 employees, is responsible for part-load and dry-cargo handling and the administration of all its Berlin locations. Its core business focuses on fresh goods services, or food organization and on dry cargo, i.e. part-loads for industrial enterprises. The Berlin branch's distribution area covers Berlin, Brandenburg and currently also Mecklenburg-Vorpommern. And it serves as the only delivery and collection area in the organization. For this distribution area, in- and outbound long-haul transports were incorporated into the national network of 40 locations. There are about 6,000 delivery customers in the Berlin urban area and about 2,000 in the surrounding area. Shippers are scattered more widely in Berlin-Brandenburg than are delivery customers, but the distribution corresponds to the 50:50 ratio between Berlin itself and the surrounding area.

Of the location factors considered relevant for Berlin's inner-city in general and Tempelhof in particular, all the survey questions were considered to be "very important". These factors are explicitly mentioned as the reason for choosing a location within the city. SPED1 prefers inner-city locations in Berlin and took an active part in the emerging discussion about IFCs arguing in favour of the inner city:

> If we were giving school grades, almost everything would get an A. All the main factors are really important. Image [...] absolutely crucial if you're in the inner city. That was my main factor when I went to the press in 94/95, when I said I'm the only big freight forwarder who's not stupid enough to move to Großbeeren. [...] That was a real novelty: a logistic centre in the middle of the city. Everyone's moving to Großbeeren, Freienbrink, wherever, and I'm staying here. Why? No-one understood why we didn't move out too, but it was a deliberate decision. It was the [emphasized] image factor that brought me into the newspapers again, that [...] put me in front again. That set me well apart from my competitors. Image, pure and simple. (Branch Manager of SPED1)

The key arguments for this categorical preference for inner-city locations are transport costs, as the company's own figures show: The fleet of 140 short-haul trucks is run by subcontractors, all of whom operate and are based in Berlin. If the freight centre were located near Berlin (IFC Berlin-South) the truck drivers would leave town for Großbeeren every morning, to check in, load up, take the B 101 into the city again, and deliver their freight; then they would collect their outbound freight, drive back to Großbeeren, unload their cargo, which would then be prepared for delivery. Then finally they would drive home again. This means that the drivers would cover the route twice more than if they were based within the city. At that time, 80 short-haul trucks were operating daily, which would have involved additional costs of about 2.5 million DM per year (appr. 1.3 million euros).

Land outside Berlin can't be that cheap, and I can't get that much state aid... I'm quicker here, I'm there right away ... when [...] or [...] sets off, I've already made three deliveries from Tempelhof. With a short-haul vehicle you make between 13 and 16 stops a day, and if you subtract the time they spend stuck in the traffic on the B 101 and driving back again, you only manage 9 or 10. If you calculate that in terms of a percentage of the short-range transportation costs, you will end up with 2.5 million [DM] for 80 short-haul trucks. In my opinion the mid-90s exodus happened because they got cheap land there, whoever sponsored it."(Branch Manager of SPED1)

Another argument against relocating is the potential labour force in the city or the possible loss of trained and qualified staff. In the mid-1990s already, and today too, the company had realized that it was more difficult to recruit trained and qualified transportation workers in the area around Berlin than in the city itself. "There's simply a higher density of workforce here, and training opportunities are better." (Branch Manager of SPED1)

The company's inner-city location has the advantage of quick access to its customers; also the staff can get to work easily. Proximity to service providers (cleaners, for instance) plays a minor role and is explicitly not considered important in respect of subcontractors (who are responsible for 98 per cent of transports). Shared use of infrastructure, cooperative services and synergistic collaboration are explicitly excluded:

No, all the synergies that we find are used to earn money within the company itself, it has a very broad range of products. If we recognize that there's a market somewhere, that a new idea might catch on, then we generally do it ourselves. (Branch manager of SPED1)

The company therefore prefers the inner city as a locale because of accessibility advantages that can be directly converted into production advantages. Inner-city synergy effects are not the reason for staying there; the sole reasons are customer proximity and production optimization. The great importance of incoming goods and the concentration of delivery customers in the City of Berlin in a ratio of 1:3 compared to the surrounding area makes this inner-city orientation including the location decision, seem only logical. The fact that other enterprises with a comparable range of services would choose a suburban location is explained as being due to two completely different concepts.

By contrast, the interviewee can understand why the big food company branches decided that as many of them as possible would move to the same (peripheral) location. Their distribution area is not just Berlin; they also have to cover fairly large radiuses. In his opinion, a further criterion in favour of the area surrounding Berlin is the fact that the food chains' wholesale warehouses have considerable space requirements. The 60,000 square metres that the previous tenant has developed at his new suburban location would definitely not have been viable in Berlin. In Berlin the gross area of sites is generally around 50,000–60,000 square metres.

In summary: The SPED1 case stands alongside two other forwarding companies which provide consolidated shipment and production-related logistics services in or for the inner city. These enterprises typically have most of their customers in the inner city and need to be located close to the market. This market proximity can be expressed in terms of space or in terms of time (via the transport network or a location close to the motorway). The best-case scenario would be a combination of both. However, SPED1 is a clear exception among the cases analysed in this branch.

Freight Forwarder 2 (SPED2): IFC Berlin-South   SPED2 has been located in Berlin for more than 100 years and moved to the IFC Großbeeren in early 2002. The company provides part-load systems forwarding and is part of a country-wide cooperation network. Integrated systems forwarding supplements parcel services, which operate in the weight class of 30–50 kg per shipment, and transports goods up to a maximum consignment weight of 2.5 tonnes. The partner network is managed and structured similarly to the parcel service companies with delivery within 24 hours and specified quality standards. A second focus is warehouse logistics. Either output is collected directly from the factory and kept in interim storage, from which all the customers' orders are handled. Or products are collected for commercial enterprises worldwide stored and delivered according to demand or sales. Complete e-commerce solutions are on the rise here.

Transport flows are mainly national; at the international level, western European sources dominate, but deliveries from eastern Europe are increasing in number. In terms of value, national transportation accounts for 60 per cent, western Europe for 35 per cent, with a "strongly increasing trend", and eastern Europe for 5 per cent, with a "very strongly increasing trend". In integrated systems forwarding, all transport parameters are specified, i.e. times, tours and routes are fixed. In Germany the network is currently serving 34–36 terminals a day; transports are not scheduled individually but follow fixed routes.

The company's previous location was in Berlin-Moabit, in Heidestrasse, next to the Hamburger and Lehrter Bahnhof (HuL). SPED2 was one of the first investors in the IFC Großbeeren and has acquired 53,000 m² of land with a developed area of 11,500 square metres. Some 260 employees work in three shifts almost round the clock. The company had long planned to relocate owing to the pending closure of its location in Berlin-Mitte and especially of the HuL container station. The previous location rated very highly in transportation terms but was no longer economically viable. Therefore the decision to relocate the business involves both push and pull factors:

> Maybe I can tell you again basically why we moved here. We had to leave the Berlin-Mitte location – which was a very attractive site, I'll tell you why right away – logical that we had to leave because the real estate prices there were just too high for a company like

ours that needs so much space, and a freight forwarding company doesn't really belong in the inner city anyway. Why would it have been attractive for us there? Transports with big trucks are mostly during the night hours, so except for a certain amount of noise the big trucks going in and out really didn't disturb the traffic in the city, not very much anyway. What's attractive for us is to break the 10 or 30 big trucks down into 80 small units, the main delivery area for this region is the Berlin city, declining steadily further away from the centre. (Managing Partner of SPED2).

The advantages of the previous inner city location are undisputed at least for parts of the company's own business. Starting distribution in the city would mean shorter distances and quicker deliveries than an out-of-town location. Especially for the parcel business and distribution services in general, an inner-city location has an anti-cyclical advantage: the short-haul trucks leave the city in the early hours and return in the evening. With an out-of-town location the trucks enter the city at the same time as the main traffic stream, making congestion worse. Policy-makers have been criticized for wanting to take heavy vehicles out of the city with no regard for traffic problems. These reservations are not considered to apply to warehouse logistics because here the traffic flows are due to long-distance and international transports or to industrial products that are delivered in full loads for interim storage and subsequent delivery to customers. This function is considered more appropriate to locations in the surrounding area. In principle, the logical consequence would be to split sites and to locate urban business in the core city and warehousing in the surrounding area.

In retrospect, it is impossible to judge whether this solution was ever considered seriously. Such a strategy would probably have its advantages but also its disadvantages with respect to operational costs and scale effects. Basically the decision to relocate to the IFC was not due solely to in-house calculations but primarily to pressure from the Berlin Senate's urban planners who had designed other development concepts for Berlin-Mitte after the political turnaround in 1989:

> We were situated right at Lehrter Bahnhof, and one day when it goes into operation, in 2006, 2008, it will be one of the city's hot spots, for everything but freight forwarding, I can understand that. Fine, that was the starting point, we had to move and then we had to decide where to? North, south, east or west? The answer was simple: to the south, everyone talks about the southern commuter belt, but that's not really so important for us. The key factor is quick access. (Managing Partner of SPED2)

Großbeeren's good accessibility offsets some of its locational disadvantages compared with the inner city, especially as time saved or lost immediately affects costs. Starting from the Berlin location and oriented towards the main delivery centres, each short-haul truck was able to deliver freight to 20 consignees, compared to only 17–18 from Großbeeren – assuming the same departure time. At first glance, the problem is comparable to that of SPED1. But the solution is different in this case; because the goods arrive early, short-range distribution can begin early too. In this way the attempt is made to avoid congestion times and to maintain the distribution productivity level as far as possible:

> Given 80 trucks, those two deliveries per truck add up to enormous disadvantages over the whole year. So we had to choose a location with optimal connections, I mean optimal in terms of traffic flows. Most of the trucks come from the south, Bavaria, [...] the old federal states, North Rhine-Westphalia, Frankfurt, a relatively long drive before they reach Berlin. With this motorway link and this infrastructure we save about half, three-quarters of an hour here, compared with Berlin-Mitte. [...] Thanks to this early arrival time, handling is finished sooner, and the small trucks can leave for delivery sooner, too. For example, it used to be 8:00 in the morning, now it's 7:30. And in this half-hour I can get the two shipments delivered, so I end up with more or less the same frequency. (Managing partner of SPED2)

This adjustment strategy does not fully offset locational disadvantage; taken in conjunction with lower real estate costs at the new location, however, the investment as a whole becomes economically attractive. This is where the trade-offs between transport costs and location costs take effect. Although the low land price charged during IFC development was mentioned as "important" (not "very important") when deciding in favour of the project, it was not considered a crucial criterion. It is more likely that the location decision was the result of a weighing-up process in the face of external pressure and hence a classic second-best alternative:

> What tipped the scales was that we had to move whether there were subsidies or not. It's logical that all the subsidies for the area surrounding Berlin were an enormous help with regard to our entire move here too, also for our future range of services; earlier I mentioned e-commerce logistics and application, topics that were included in our planning although they're still in their infancy. Seen today, if things had been different with regard to the subsidies, we'd have had a leaner version of the move, no way about it. But the decision whether we would have to leave in the end, that was clear. No, the tax benefits for businesses here are offset a lot, or a bit, by the better inner city location and the costs involved. (Managing Partner of SPED2)

In summary: The SPED2 case is in line with a series of many interviewed freight forwarders who mainly operate commerce- or production-related warehousing and logistic services that do not necessarily need to be located near to Berlin or its metropolitan area, but who do some of their business in Berlin-Brandenburg. Of the cases analysed, this case clearly represents the majority; in other words, it is the rule.

Retail Distribution LM1: Berlin, Inner-City Southwest The firm operates a grocery retail chain for about 100 years, initially in cities all over Germany. Today, it is still present in four agglomerations, including the Berlin-Brandenburg region. In this region, the firm owns about 160 retail outlets, mainly in the core city and in selected parts of the surrounding area (e.g. Potsdam, Zossen). Three delivery warehouses had been operated until the year 2005 on three different sites in the Mariendorf-disctrict in the southwest of Berlin. The gross size of the areas comprised about 100,000 square metres. Also, a home-delivery service is being offered, organized from the company's site of the *Berliner Großmarkt*. By the end of 2005, a new, central warehouse opened up in the Mariendorf-district on a former gasworks industrial plant. The firm is situated in the southwest of Berlin since the 1970s. In its own words, it is embedded into this local framework, particularly due to the local labour market. For this reason, once the new facilities had been planned, there was no location outside the district that had been taken into account.

Two of the three old facilities were dedicated to dry commodities storage, the third comprised a warehouse for perishables. Dairy products had been organized by an external broker who was responsible both for the commissioning of the consignments and their delivery. This business is supposed to move back to the core business of the retailer. A second external broker organizes the refrigerated food services, located in the southeast of the Berlin region, which is also discussed to be pulled back. This kind of "insourcing" appears less exceptional than it looks like:

> We are going to do this because fruit logistics is quite similar to the dairy distribution, and it is worth to become bundled. The delivery rhythms also seem to be similar. Fruits and vegetables e.g. require daily delivery; dairy depends upon the expiration date. These are not as critical, however, but you have to go to the outlet in any way, so you may be able to drop costs. The regular commissioning for the dry commodities reflects our aim to keep the stock as minimal as possible. Since delivery distances are relatively low, we prefer to keep the stock in the warehouse, rather than in the outlets. This offers us a higher degree of control, and we avoid an overstock in the outlets. This turns out to generate higher delivery frequencies, compared to our competitors. Each outlet receives deliveries three or four times a week, which appears to be relatively frequent. (LM1: Logistics Manager)

LM1 operates its own fleet, yet in the case of additional demand also calls for freight forwarders. This applies particularly for peak demand hours, which makes the ratio between own and third party transport being 50:50. The current number of trucks that are being handled does not exceed 200 a day with a minimum of 10 tons of load capacity. The average load per vehicle is about 20 tons. The dry commodities segment moves about 200,000 tons per year, the fruit segment about 70,000 tons, dairy products about 35,000 tons per year. Certain direct deliveries have to be added as well, e.g. regarding bread or non-food. In this case, the delivery includes the staffing of the shelves and reverse logistics (e.g. of perishables). Finally, the retailer also operates a meat factory in the western suburbs of Berlin that provides all outlets in the region with meat etc.

According to a frequent pattern of logistics reorganization, the retailer has consolidated the three existing warehouses and concentrated all operations in one – urban – warehouse. This results in reduced overhead-costs and a more efficient logistics operation. In so doing, the firm finishes a historical process of concentration, since in the old Berlin (West) the urban area did not offer appropriate sites for distribution firms, particularly in terms of size. Before, the goods distribution industry had significant problems related to land rents. The basic idea of Berlin policy and planning was then to offer appropriate sites to distribution firms, particularly by establishing suburban freight centres. LM1 however experienced a sharp restructuring by the end of the 1990s, which ended up in a significant retreat from the rest of the New Länder (eastern Germany), thus clearly focussing on a core city location for warehousing and distribution purposes.

> The experience of operating the meat factory revealed that we need to stay close to the city, in order to provide our mainly urban outlets. Land rents and the price for lots had not been decisive in this respect, but so were the long-term transport costs. We have appreciated to learn that all our competitors had been moving to the suburban Integrated Freight Centres, either in Freienbrink or in Großbeeren. I am always keen to learn from a particular competitor who is quite well organized in terms of logistics. They'd been moving to Großbeeren. In terms of distribution areas, they had been dividing Berlin into two areas: north and south. [...] After having moved to Großbeeren, they still operate a north-south divide, and this may make sense in their particular case. Others who have their outlets mainly in the Berlin area and have moved as well: they are now paying off. (Logistics manager, LM1)

The decision for finding a location in more or less the same district as before was also "political": Since the majority of customers (and outlets) is placed in Berlin, the firm declares its interest in creating jobs for Berlin as well. This would also be an outcome of social responsibility, even against the background of the failure of Berlin economic policy (which is considered inflexible from the businesses' perspective). Proximity to customers and the local embeddedness of the firm were the main rationales for the decision to remain in Berlin and not to move into the suburbs, even not into an Integrated Freight Centre (IFC):

> Our rationale was clear: we'll never chose an IFC for location. Just look at the traffic bottleneck. I don't want to go where many big companies, particularly freight forwarders and logistics firms, go to. The result is congestion, both derived from incoming and outgoing deliveries. I always want to be a lonely settler, so I won't face serious transport problems. (LM1)

The decision for chosing a location in the south of the city was not only made due to the location of the former three warehouses, yet, in the context of previous plans for a joint logistics organization (and concept) with a discounter that belongs to the same corporation. This model would have been based on a combination of discount- and supermarket distribution. However, the presumable partner preferred to operate his logistics organization independently (and had moved his warehouse to the northern beltway, just at the city border). Against this background, a southern location would make particular sense. The main factors that supported the southern area were transport costs due to the direct access to the majority of the outlets. Also, any changes for the employees remained limited: LM1 could transfer mostly all previous staff into the new facility. This would be explained by qualification aspects as well:

> It is some time ago that logistics mainly employed less qualified people [...]. Many of them have to do with PCs and computerized systems, hence we are glad if we can hire qualified workforce ... It is also important in terms of trust: once we now each other for long, you can count on your staff people. If you're going to hire new staff, you need some time for getting about who would fit for long-term engagement. And in this respect, this location turns out to be ideal. (Logistics Manager, LM1)

However, looking at the locational profile in detail, this particular place ranks "second best" only, compared with the *Berliner Großmarkt*, where LM1 operates its home-delivery service for private customers:

> The best place for a warehouse would be the Beusselstraße. It is close by the Stadtautobahn ... This is where I would like to go. But such places are no longer available there. For organizing the home-delivery service, this is a perfect place: in the heart of the city, good access, with the Stadtautobahn close by, good to get there and to get off. The facility is our own, and the home-delivery service is well established on the market. (Logistics Manager, LM1)

In order to summarize this case: LM1 is quite exceptional among the retail chains that are serving the Berlin-Brandenburg market. It is the only warehouse location that is operated at a more or less inner-city location. All major competitors have been moving either to an IFC-location or to other suburban commercial areas. The main reason for this locational policy is access to customers, since the large majority of retail outlets of LM1 are located in the Berlin urban area; in addition, just a few outlets exist in surrounding areas. Especially transport costs have come into play in this respect, which were more important than land rents or real estate prices, respectively. LM1 also noted to be a "Berlin corporation", particularly interested in the city, also in offering jobs for the local community. With this statement as well, LM1 appears exceptional among grocery retail chains of that particular size in general.

Retail Distribution LM2: IFC Berlin-South   LM2 operates a new distribution centre in the Integrated Freight Centre (IFC) Berlin-South which opened up in August 2002. By operating the new warehouse, the company could shut down three single locations: a refrig distribution facility in the City of Brandenburg/ Havel, the dry commodities segment based at Berlin-Lichtenberg and an inner-city DC for fruits and perishables distribution. These three warehouses had been consolidated on one site of 12.5 hectares. The main building comprises a space of about 42,000 square metres under roof. The facility employs a total of 435 people. The DC is one out of 23 facilities operated by LM2 nationwide and is serving about 350 retail outlets. The distribution area stretches until the Baltic Sea shore (Rügen), until Sachsen-Anhalt in the west and Saxony in the southeast, thus serving the "Länder" Mecklenburg-Vorpommern, Sachsen-Anhalt, Berlin and Brandenburg. The West of the City of Magdeburg, the next distribution area is designated which is being served from a DC in the Hannover region; a second location for distribution in the area of the New Länder is in Saxony. However, about 60 to 70 per cent of the outlets served by the IFC Berlin-South are located in the Berlin-Brandenburg region.

The freight distribution is operated by freight forwarding firms and specialized service providers, both in the dairy and fruit segment as well as with regard to dry commodities. The service providers aim at bundling effects, in order to deliver complete consignments to LM2. Thus it is intended to minimize the incoming delivery trips and to synchronize the time of delivery. Regarding the outgoing distribution, freight forwarding vessels are being operated, in a broad range from solo vehicle up to trailers. A total of 110 vehicles provide deliveries for the outlets, operating an average of two tours per day, with a load of 3,500 to 4,000 pallets. This results in about 400 vehicle movements at the location of the DC per day. Transport services are exclusively operated by third-party providers, solely using trucks. Other transport modes are not in operation due to time constraints and flexibility requirements.

The locational choice in the case of LM2, which in the end favoured a site in the IFC Berlin-South, was preceded by a screening procedure that included the Berlin-Brandenburg region as a whole, and also different scenarios for designing the optimal distribution area. Regarding the different models favoured by the competitors, the challenge always remains the same: how to ensure the best access to the core urban area? Given the next distribution centre located in Saxony, the Berlin-Brandenburg macro-region was soon to be chosen. Afterwards, several locations had been investigated especially at those locations where building permits could be retrieved quickly from the municipalities, so that the project could soon become realized. In the case that a potential second distribution centre would become necessary in order to serve the Berlin-Brandenburg outlets, a location close to the northern Berlin beltway would then be favoured.

Among the locational factors highlighted by LM2, the size and particularly the price of a lot were not valued as high as it is oftenly practised. In contrast, the functionality of the traffic system and particularly motorway access score highest. In this context, this criteria is not only related to traffic operability but also to the acceptability of the neighbourhood. LM2 requires a general industrial designation of an area that in principle allows for 24/7–operations:

> Our shifts are running from Sunday mornings at 9 until Saturday night at 11. The related vehicle turnover runs from Sunday night at 10 until Saturday noon. Well, this is a major,

decisive criteria. If you are constrained by too many zoning regulations, buffering and design policies etc, this is not operational. Although our noise emissions are quite limited compared to the new trunk road that runs across the IFC, however, the refrig aggregates on the roof, they are essential. (Logistics manager, LM2)

Whereas the IFC label as such (multimodal infrastructure, corporate network of freight firms, image) did not play a major role for site selection, this decision had been made with respect to the infrastructure that is provided here in general. On one hand, motorway access is ultimately required and also ranks highest in the evaluation of all micro-locations. Regarding the direct connection between the IFC and the Berlin motorway A 10, this particular corridor is well established. The main advantages of accessibility provided by the motorway network come into play once designing the location. A major problem was once derived by the missing link between IFC and Berlin (the B 101 trunk road), which meanwhile had been completed.

Despite a general satisfaction with the IFC location, the warehouse is about to exceed its capacity as a consequence of a cost-sensitive planning and a conflict between logistics on one hand and controlling on the other, which is supposed to emerge as a generic contradiction in corporate policy:

There is a simple reason for conflict, since ... initially, a concept had been developed that raises questions either from the logistics or from the cost perspective. Regarding logistics functionality, a lot that includes parking spaces for about 50 trucks would be desirable as a buffer zone, usually framed around the buildings. From the cost perspective, this request was considered too expensive, even if the business would successfully develop: compared to the 42,000 square metres floor area, 50,000 would have been even better. This would allow to expand the range of commodities supplied and also to operate in a more flexible way. The same applies to the full integration of different warehousing sections, such as meat, that had not been considered to be managed in a general DC before. (Logistics Manager, LM2)

The previously owned locations in Berlin had been closed down due to neighbourhood conflicts caused by traffic related noise emissions. The resulting regulations would not have permitted early morning operations of incoming trucks. Basically, the location decision that favoured the suburban site would be made in a similar way again. However, given this case, LM2 would then select a location close to the northern Berlin beltway, before complementing in the southern area, since the next DC is in operation in the south of the New Länder, from which the southern part of Brandenburg would have been provided as well.

In summarizing this case: LM2 represents the usual way of placing grocery distribution centres either in suburban areas or even in more remote places with good traffic access, predominantly with regard to the road transport mode. Inner-city locations are generally out of consideration for this particular purpose, since they do not provide the required size and operability, and yet, often also include critical neighbourhood and community effects. Different from freight forwarders, the management of the grocery commodity chain appears to be easier, since the own outlets ensure a higher delivery density than the service of scattered consumer locales does. Therefore, the number of DCs operated nationwide (23) is far below most of those of the freight forwarders (about 40). In the case of LM2, the fewer number of nodes allows for a more efficient, "economies of scale-oriented" provision of even larger distribution areas.

Network Building and Spatial Orientation

Are there Logistics-"Networks"?

A major question raised by this research was directed to the assumption that logistics would generate particular network structures and thus determine the locational behaviour of those firms that would require or at least prefer proximity to freight distribution services. Particularly the local networking processes that are conceptualized in the theoretical framework of "geographical industrialization" could play a key role for the assumed urbanization effects and thus deliver empirical evidence to the research approach. With particular reference to the emergence of networks and relationships in economic geography in general and location theory in particular, logistics is considered a key issue in maintaining the organization of economic networks.

However, due to the almost ubiquitous supply of infrastructure, not least thanks to the modernization of transport networks after 1989/1990, Berlin-Brandenburg as well as other economic areas of the New Länder are embedded in the particular spatial division of labour at the national and increasingly also at the international level. In this regard, large-scale economic relationships between corporations are predominant with no specific preference for certain areas. Also, there is no particular regional bias, e.g. as a consequence of a re-emphasis of the region as an economic space. The only exception may be the somehow path-dependent strong links between Berlin and West Germany that originated in post-War times. This historically determined and extensively stretched hinterland of Berlin is still at work today. Wholesale traders are used to organize their supply from industries in North Rhine-Westphalia or Baden-Württemberg. These economic relationships may be changing over the coming years, being replaced by a more regional basis of interaction.

The logistics corporations investigated by this study are basically connected with customers all over Germany and also Europe, both regarding incoming and outgoing consignments. Their distribution networks aim at covering all major economic regions without having regional preferences. The same applies to the business of grocery distribution, where purchasing is centrally organized and normally determined by aspects such as quality of products, price and the standard of delivery service. Geography in terms of regional sourcing does not play a significant role with a few exceptions (e.g. in the organic grocery market). However, specialized service firms providing certain commodities do have a distinct regional supply base; this is e.g. the case with Eastern Europe that is extremely important for the wholesale trade with game or poultry. This business is even today pre-dominated by particular institutional relations, mainly to exporters that were already powerful before 1990.

Regarding the main research question of this study, the issue of locational patterns and dynamics of distribution firms, this is of less importance. The large-scale economic exchange which is either widely spread or still maintained with firms in the Old Länder has influenced the regional land use patterns in one way only: it has reinforced the strong preference for the southern Berlin beltway as a sub-regional location, rather than the related commercial areas in the northern or eastern part of the study area.

Against the background of the search for regularities in the locational patterns and respective dynamics of distribution firms, this study was focussing on the regional linkages and interactions between the corporations. This research question ties up to two threads of discussion: on one hand, the issue is about regional networks, as mentioned above; on the other hand, freight centres are considered as particular means of agglomeration, both for similar and different types of corporations. Its supposedly positive impact on emerging network relations was a major rationale and legitimation of IFC as a product and subject of massive public support. The freight centres were considered to generate "critical mass" by concentrating certain firms, thus making modal shifts from road to rail transport economically efficient or possible at all. Also, as prime logistical nodes, they should assist in organizing collective politics of "city-logistics"; the associated optimization of delivery transports into the core city area also requires a critical mass of consignments and thus participating firms. Such synergies would also help to further promote the idea of integrated freight centres in general, once gas-stations, hotels, tire-dealers, truck-services and repair or even sub-contractors are located on the same site.

In this particular respect, the three Berlin-Brandenburg IFC did not meet their own standards that were put forward more than ten years ago. Despite a few exceptions, it is not clear whether such linkages and inter-relations among the firms would in fact be substantially practised. The large majority of distribution firms investigated, regardless whether they offer freight forwarding and logistics, grocery retail distribution, wholesale trade or courier services, represent more or less isolated entities. They do not pursue strong links to other firms that are located in the same commercial area – despite the fact that sometimes freight forwarders and shippers let regional transport firms take over the task of physical distribution. One representative of a parcel and courier service even mentioned that the location choice for placing depots would, among other factors, be made in relation to the availability of subcontractors who carry out the delivery. However, there is no convincing evidence supporting this argument. All major corporations of this subsector mentioned that they may find subcontractors anywhere. Also, the subcontractors would follow the lead logistics provider and freight forwarder in the case of locational mobility – not vice versa.

The survey among firms that are located in the Integrated Freight Centres revealed that there is no reason to assume that, first, certain linkages play an important role for decisions made in terms of site selection (see Figure 4.3); it is also not likely to assume that such linkages are now being practised among these firms (see Figure 4.4). 66 per cent of the surveyed firms responded to the question whether such linkages exist in their practice with "no", 4.3 per cent of the firms do operate such relationships at least punctually. 19.1 per cent of the firms suggest that this aspect had played a role once they were seeking for location, whereas 10.6 per cent responded "punctual". 55.3 per cent did not confirm the existence of such linkages. In this respect, freight forwarding and logistics firms do not distinguish from retail grocery distributors or from wholesale traders.

What could explain this low degree of inter-firm relationships and linkage? According to the survey, there is one convincing explanation: modern logistics systems have undergone a significant organizational and technological change over the last decades. This modernization has mainly aimed at optimizing single firms' operations. Inter-firm activity or co-operation have not been pursued, except one model: the bundling of small-scale, cost-intensive consignments by freight forwarders that have put together and thus harmonized their networks, have shared distribution areas and thus aimed at improving their joint competitive position. On the basis of the relatively high efficiency of such operations and their defined, high quality standards, these services can become definitely improved.

Moreover, the single firm appears to be integrated or embedded vertically according to the structure of the value chains. However, the organization of these chains usually occurs in quite a rigid way, top-down oriented and is not open for joint action at the horizontal level. The broad literature on the geographies of commodity chains (see Chapter 3) revealed the strong influence that originates from power distribution exerted on the chains. This applies, first, to large firms that orchestrate voluminous flows and concentrate a high amount of purchasing power. Second, some firms may occupy key positions in the chain, such as large shippers or large freight forwarders that are able to determinate logistics management, down the stream of the supply chain. Also, the related IT-systems are oftenly not compatible with those of other firms, which sometimes can be considered a particular competitive advantage. Technical standards may play a role and hinder inter-firm co-operation, both regarding e.g. different corporate IT-solutions or the requirements of different transport modes.

Horizontal co-operation among distribution firms seems to be the exception rather than the norm, based on this study. IFC-firms "co-operate" once they are buying electric power at discount fares on the deregulated market, they jointly organize road maintenance within the IFC, and they are concerned about improving the image of their location. However, further impacts seem to be limited, since the logistics organization of a single firm is usually distracted from the organization of other firms' logistics. As a consequence, a majority of respondents do not

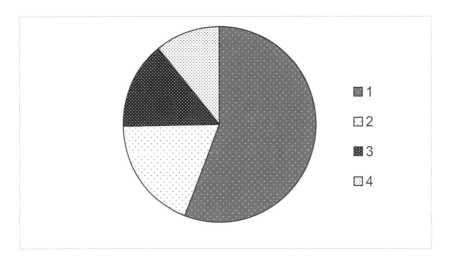

Figure 4.3 **Were there any network relationships important for your site selection?**

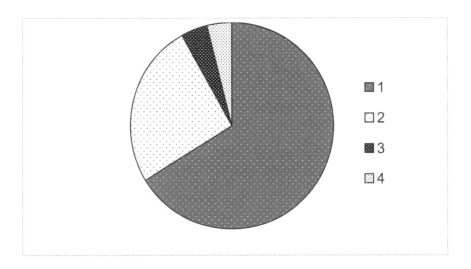

Figure 4.4 **Are there any local network relationships that you are practising right now?**

favour the idea of networks for distribution firms. Even avoiding proximity to other companies of a comparable size is sometimes desired explicitly, as the retail distributor LM1 mentioned above exemplified, who does not want to be exposed to the disadvantages of agglomeration, particularly congestion and emissions.

LM2 however is positively concerned with potential network effects and practices linkage to firms at the IFC-location. Although activities based on such linkages only represent a minor portion of the total amount of inter-firm relationships: the IFC as a place of networking has influenced the site-selection decision of the firm.

> Well, we thought o.k., if there is an infrastructure emerging, then it might be wrong to chose a totally different place. In the New Länder, many municipalities have designated commercial and industrial areas where there is virtually nothing. If you look at this particular agglomeration, there are many other firms of the same business, all competitors. If there is a concept as such, which particularly aims at these kinds of firms, then it is simply reasonable to go there. (LM2: Logistics Manager)

The example of the IFC Berlin-South reveals the degree of inter-firm-relationships and networking in a typical way: the freight centre is well developed, almost fully occupied and it hosts representatives of the most important components of the supply chain, being this major shippers (i.e. retail chains) or transport service providers (freight forwarders, carriers). However, even this case of a relatively advanced freight centre documents that, even if logistics firms with a similar or different profile are situated in a common setting, this must not necessarily mean that as a consequence, a "network" or a "centre" emerges bringing about particular impacts such as synergies etc. Joint activities represent also in this case the exception, rather than the norm:

> Well, there is a freight forwarder just across the street [...], with which we are co-operating. Once we've called them and have talked to them, yet this is the only case where we are doing that. [...] The freight forwarder over there, he had also been asking for possible co-operation; however, he's mainly offering LTL-services and does not provide the kind of truck that we need (with special features such as refrig aggregates). In contrast, the other freight forwarder is already under contract in other German regions. They simply developed a new service line according to our needs, including new vehicles ... Moreover, we are strongly connected to other service providers that are located in the IFC-Berlin East. The reason for this is that they simply started their regional business from that particular location. Now they are operating practically a new subsidiary on our site. (LM2: Logistics Manager).

Isolated locations may often exclude such co-operations per se, so there will be no further reason to seek for networks etc. However, as it has been widely practised in the past e.g. in dispersed, rural areas, the shared use of expensive machinery or service providers can be put forward even among partners that are territorially more distanciated. Of course, physical proximity or neighbourhood is no strict requirement for making use of eventual linkages and co-operation. So the idea of the network could be extended to commercial areas that are located in the same region. This would allow for the joint use of vehicles, machinery or even services:

> Well, there are repair services provided in the eastern leg of our distribution area, rather than in the western part. The western part is a fairly lonely setting ... There is nothing but our parcel centre. There is the village, and there is our facility, anything else. In contrast, in the eastern district there is much more provided, there are repair services, there are freight forwarders close by with whom one *could* do something in practice. In the western district, we are organizing vehicle operations for a freight forwarder close by ... this means that we are sharing trailers for on-site movements of vessels and containers. The main reason for this is that the respective time windows are appropriate – the capacities were complementary required by each firm. Were are practising that as far as possible. As long as we were operating a long-distance parcel-express train, we were co-operating with a private rail freight carrier. They were managing the local in- and outbound freight movements. We are doing this, as far as it is possible. In our western district, it is a bit more sophisticated and developed than in our western district. (KEP2: District representative).

A wholesale trader who operates his distribution centre in the southern surrounding area of Berlin once pursued the idea of a particular centre for sanitary technologies and equipment. He had related firms in his immediate vicinity that would have fitted in this joint centre. However, the idea could not be realized due to a periodically difficult business climate:

> No, we don't do that. Once I'd tried to establish something like that. A kind of technical centre for home and sanitary equipment. We've had some suppliers over here which could have perfectly complemented such centre. They were operating their own small warehouses for storing their specialized stuff. Then we told them that we are a wholesaler, committed to operate a full-range distribution centre. It would be much more efficient and cheaper for everybody over here to pull our commodity stock together and deliver out of the DC. This was, by the way, the reason for many wholesale trade firms to leave the core city of Berlin: the lack of affordable, efficient warehousing space. Before 1989/1990, due to the scarcity of such resources, many wholesale and trade firms pursued such co-operations. Thus they could significantly drop their costs. After having moved to their suburban outlets, there seems to be no good reason for sharing warehouse space. This is actually stupid, because we all could drop our costs. (Executive WHO2).

To some extent, networks may even emerge by coincidence, if they are not being considered in the early stages of site selection. This is demonstrated by a medium-sized freight forwarder in Potsdam, who not only operates his own business, yet, also developed a small local cluster of service firms, some of them among his customers:

We are already practising co-operation with firms that are located in this area. This happens to the benefit of everybody. We've got several vehicle-repair shops over here. At short distance for getting trucks fixed etc. Our own firm is even active in the real estate business, particularly of commercial space. There are several wholesale traders and distributors that we could attract to mover over here, and they are also among our customers. So there are indeed some linkages. This has all been developed by ourselves. It is not completed yet, however. I was always convinced that we need more customers for this area, particularly those for whom we could also offer freight forwarding services. And this is indeed the case. This has worked out quite well up to now. We have developed an office building over here, there is no tenant in there who would not be our customer in terms of warehousing. These are mostly small firms who need a small portion of office space, and they also need about 200 square metres warehousing space, or 100 square metres or even 500 square metres, respectively. We are able to meet their demand due to our warehousing services. And there are also consignments to be handled. Among our tenants, there is a wholesale trader in medical technologies. He needs about five pallets to be handled a day. On this basis, he could never afford to buy a forklift. Instead, our services are quite efficient in this respect. (Executive SPED3).

So the Executive had also been active as a developer of the site from which he operates his freight forwarding business. In addition, he could successfully attract 27 small and medium sized firms to chose that particular site for location. The benefits of such development activity are clear: he is able to lease office and warehousing space he does not need on his own. This case of a small commercial park corresponds with a so-called "private IFC" in Ludwigsfelde south of Berlin, where the main user of a lot of about 2 hectares has located other related firms. These are both customers and complementary services, even a customs office is on-site, due to the high share of consignments that are handled for customers in eastern Europe. These two cases resemble the general idea of freight networks that were pursued with the establishment of the relatively large-scale IFC in the surrounding area of Berlin. They appear more in the shape of mixed-use developments that are primarily initiated in core urban areas, rather than a freight centre in the sense of this term. Usually, they are established to provide small enterprises, such as craftsmen etc. with space for operation and also with additional services, e.g. shared office functions. The marketing of these concepts is practised on a small scale, and in most cases, they are completely privately financed. This stands in remarkable contrast to the extensive marketing with which the IFC had been promoted, not to speak of massive public investments that had been spent for infrastructure and also the subsidies that make the land rents competitive at IFC-locations.

The marketing practice of the Potsdam commercial area introduced above appears both strategic and also improvised:

We have placed some signs which are visible from the trunk road that is quite frequently used. Our firm has a good image in this region; our salespersons are very close to the customers; I am also on the road rather permanently. There is often the case that

I am asked for possible co-operation or for complementary service providers, and I may be able to help out. And, also, there is a lot of coincidence. Across our building, there is a medium sized wholesale trader, based in Berlin, who operates a distribution warehouse. Once I was about to leave my office, there was a black Mercedes car from Berlin coming, three business people in there, interested in our area. Well, I stopped and asked whether I could be of assistance, and we chatted a bit. As a consequence, we've made a contract, afterwards the wholesaler constructed a new building on our site. This deal already runs for about 5 years, and will hopefully last for the coming 10 to 15 years. [...] We are also quite well in touch with the Chamber of Commerce and the City of Potsdam, without having received any subsidies yet. They know our concept, and there are certainly calls whether we might have space for firms or could fix a problem. (Executive, SPED3)

Spatial Orientation of Logistics

Core city and suburbs in Berlin have been developing in a different way regarding the locational dynamics of logistics and freight distribution. Insofar, the main hypothesis suggesting a significant suburbanization of distribution appears to be confirmed by empirical evidence. Whereas the number of employees in freight transport, freight forwarding and warehousing (except wholesale trade) has been decreasing in Berlin by about 30 per cent between 1993 und 2005, the related number has more than doubled in the municipalities of the surrounding area (see above in more detail). Since the year 2001, for the first time since 1989/1990 distribution employment located in the suburbs was higher than it was in the core urban area of Berlin. With respect to the distribution function, the suburbs are now more important than the city is, although the population in Berlin is three times higher compared to the suburbs and occupation is twice as high.

The assumption of a significant suburbanization of distribution is being confirmed regarding the development of spaces devoted to freight handling and warehousing. According to estimations carried out by real estate firms, the building block of about 6 million square metres warehousing space in the entire Berlin-Brandenburg region is almost evenly distributed (50:50) between city and suburbs (JonesLangLaSalle 2001). A careful interpretation of this data has to take into account that these numbers include the warehousing space provided for or by manufacturing companies as well, so the findings may be biased to some extent. However, the data indicates that the suburbs have been performing disproportionately well in this particular respect. Judging from an older assessment of the functional orientation of commercial areas in the surrounding areas of Berlin, the sector of logistics and freight distribution comprised about 21 per cent of all firms and thus represented the largest group of suburban commercial land uses in that area (Aengevelt Research 1999, 21). It seems to be evident that suburbia became a "logistics organization space" (see Chapter 1).

Regarding the question of spatial orientation, it is much more difficult to make a similar statement on goods flows. This is mainly due to the lack of data, since the official statistics do not (and will never) cover the amount and direction of corporate freight flows, at least in a regional breakdown. Also, there is no clear distinction between those flows that originate or end up in the suburbs and those that are oriented elsewhere. In the year 1999 (there is unfortunately no recent data available on this), companies located in the three IFC have handled an amount of 4.7 million tons of road freight consignments, in addition also 5,000 tons of rail freight (Landtag Brandenburg 2000). Judging from the total volume of freight flows generated in the two Länder Berlin and Brandenburg of about 92.7 million tons, the share of the IFC related handlings was quite minimal and did not exceed the 5 per cent margin. A proper comparison, however, is actually not possible on the basis of these data.

Despite this contention, it is evident that the suburbs are on the way to become the terminal for handling the freight flows that are directed to and originate in Berlin. How can the related role of the suburbs be assessed in this respect? Is the suburban area subordinated in terms of freight distribution, since much of the flows are connected with the city area not the suburbs on their own? Or do the commercial areas support a more independent development, a kind of emancipation of the suburbs from the core city? The findings of this study indicate that, first, to a larger part the suburbanization of logistics still consists of the movement of firms out of the core city rather than of the establishment of totally new developments. This is more in line with a traditional interpretation of sub-urbanization, rather than of a profoundly new role of the suburbs in the context of a larger, more complex urban region.

> There were predictions and, even more, hopes for firms coming from elsewhere outside the region. However, the major part was related to the suburban drift originating in Berlin. Even today, if there are anymore movements going on at all. (Researcher REAL ESTATE3)

The research question that was raised with regard to the meaning and typology of suburban development can now be discussed in more detail, based on the response given by the distribution firms suggesting to what extent their business may be confined to the core city; and what portion may be more related to the suburbs. The latter can be interpreted as an indicator for the tendency of a more independent development of the suburban areas.

The role of Berlin (and that of the suburbs, respectively) can be judged from the share of freight distribution that is related to each of these different spatial categories. During the personal survey, the firm representatives could respond in a range of five scores; also, there was space for a deeper discussion of this issue during the interview. In order to summarize the findings, first, a major part of the corporate activity has indeed been related to Berlin. This applies to Berlin firms per se, yet for suburban firms as well. 57.4 per cent of the firms surveyed had classified their spatial orientation as "very strongly" related to Berlin, 21.3 per cent voted in terms of "strongly related". 8.5 per cent of the firms had argued as being "less strongly" related to Berlin. Only 6.4 per cent of firms were not specifically related to Berlin. Accordingly, almost 80 per cent of the coporate activity covered by this research was related to Berlin – the respective distribution is more in line with the general total population and occupation relation between both categories. In contrast, the locational setting of logistics favours the suburban places.

This result comes to a surprise on one hand, since the corporate survey covers several firms that operate widely spanned supply networks reaching from the south of the New Länder up to the Baltic Sea shore. On the other hand, it is of course an outcome of the partly extreme disparity between the densely populated core city and the relatively weak, scattered suburban part of the region. It is indeed the network configuration of each corporation that mediates the degree of spatial coverage and the related number of hubs and nodes.

The broad range of the related attachment to Berlin can be exemplified in the case of some of the firms introduced earlier. The LTL-freight forwarder has his main customer basis in the core city. Consequently, the main business is being operated in that particular area. Being asked whether he would move to the suburbs, he clearly denied: "In the city, for the city – that was always our label and header." (SPED1: Logistics Manager) The same applies for the freight forwarder who is located in the suburb, yet with a majority of his customers remaining in the city, and a significantly lower share in the suburbs or in the rest of Brandenburg: "Two thirds in Berlin, one third in the suburbs." (SPED2: Executive)

The "very strong" orientation of the first of the grocery retail distributors to Berlin is a consequence of history, corporate restructuring and concomitantly location as well (LM1, Logistics Manager), whereas LM2 appears differentiated in this respect. The suburban, IFC-placed DC provides a much larger area compared to the suburban zone and even to Berlin. However, judging from the number of outlets, "60 to 70 per cent of the outlets are in Berlin" (LM 2, Logistics Manager), which is mainly derived from population density. Two wholesale traders that had been investigated accordingly confirm this view: a machine tools wholesale trader located in the core city area has about "90 per cent of the customers ... in the Berlin-Brandenburg region, and of that 40 per cent belong to the core city" (GH1, Executive). The related wholesale firm that is located in the suburbs deploys a less-extensive Berlin orientation, as one would expect. However, this picture is also changing, since the State of Brandenburg is badly performing economically, so there is a shift underway back to Berlin:

> Well, Berlin now comprises about 60 per cent, since the market here in Brandenburg and on the edges has almost totally broke down. In earlier years, it was about 50:50, even though 55 per cent Berlin […], it is shifting more towards Berlin, the initial market demand has been served, whereas the suburbs are significantly weak. There is a particular lack of industries, and this will remain in the coming years. (WHO2)

The courier and express service firm KEP1 performs a strong Berlin-orientation, yet, has divided the distribution area almost evenly between a western and eastern district, both including parts of the core city and the related suburbs.

> Yes, there is a certain amount of the incoming volume that is directed to the suburbs, this is a consequence of the population density, so there remains a significant portion for Brandenburg. In the opposite way, regarding the outgoing consignments, a majority originates in Berlin by far. In this respect, Brandenburg has remained fairly poor, particularly in its more distant parts. The bulk of our outgoing parcels, I suggest 60 per cent of the consignments, are coming from Berlin, 40 per cent from the suburbs. Regarding the incoming flows, it is about fifty-fifty. (KEP2)

Among the freight forwarders that are localized in more remote areas of the State of Brandenburg with DC-functions related to contract logistics, the orientation to Berlin may be even lower. In this case, Brandenburg is one component within the national network and comprises a minor share only, which is of course related to the low population density and the also low industrial basis after years of transformation and de-industrialization. Within this context, Berlin occupies a certain portion, yet, does not necessarily dominate the business. One contract logistics firm that locates close to the southern beltway, the Germany-wide business accounts for about 75 per cent, the Berlin-Brandenburg region for about 25 per cent; 60 per cent of this quarter is related to Berlin. A grocery freight forwarder that has settled on the edge of the study area, somewhat remote even from the motorway, operates about 65 to 70 per cent of his business volume for customers in Berlin. As far as the entire Berlin-Brandenburg region is to be covered, the core city of Berlin plays a central role in this respect.

In order to summarize this issue: customer structure, spatial distribution and also the logistical concept may determine the degree of core city orientation of the distribution firms. They respond to this regional framework by developing a differentiated locational setting: with either central or decentral location concepts depending on the area to be covered, or by developing multi-location concepts, in order to cover both core city and suburbs (or even parts of that). All firms however are more or less significant in their factual orientation towards Berlin. Even the nationwide distribution concepts, operating single locations in the region (which actually do not make sense, given the somehow peripheral location of Berlin-Brandenburg in relation to the rest of the country), are to a certain degree related to the Berlin-Brandenburg region. The "logistical satellite" that might be distracted from the region and its core is hardly to identify. The same applies for the strategic "Brückenkopf" that connects the region with Eastern Europe: such function could not be identified in this study. Also, there is limited empirical evidence supporting the view that the suburban firms may cover the suburban area exclusively. They are part of wider supply- and distribution-networks of the region and thus specifically adjusted to the entire region and its different subareas.

Chapter 5
The Northern California Case Study

The San Francisco Bay Area/The Central Valley, California/USA

Structure, Shape and Dynamics of the Region

The second case study has been conducted in the San Francisco Bay Area in Northern California, particularly in the East Bay Area and in the Central Valley. The San Francisco Bay Area represents one of the most dynamically growing regions of the U.S. Particularly after World War II, the Bay Area has undergone a significant transformation in the context of urbanization and industrialization with the emergence of the high technology-district in the so-called Silicon Valley in Santa Clara County, the southern area of the region as one of the most important developments. As the S.F.-Oakland-San Jose Metropolitan Area, it represents the fifth largest metro area of the U.S. with a size of almost 18,000 square kilometres and a population of almost 7 million by July 2006 and 3.4 million employees by May 2007 (data after U.S. Census Bureau 2007, California Employment Development Department 2007) (see Map 5.1).

Urban development in this region has occurred according to spatial cycles with rapid industrialization starting in the second half of the nineteenth century. In retrospective, Walker (2001) has interpreted such cycles by using the frame "industry builds the city": spatial development has been considered to be mainly influenced by the locational dynamics of commerce and industry. The case of the Bay Area was not a case of an industrial core that has been subject to suburbanization later on, yet, suburban expansion beyond existing borders has been the normalcy of regional development (ibid., 53). As a major factor of influence, Walker mentions land rents and speculation, an improved infrastructure, industrial relations, regulation of labour and innovation.

In terms of population, the S.F.- Oakland-San Jose-Metropolitan Area has deployed the second highest growth rate nationwide between 1990 and 2000 (12.6 per cent), only exceeded by the Washington-Baltimore Metropolitan Region (U.S. Census Bureau 2001). By judging the region's gross domestic product per capita, the Bay Area ranks first among all U.S. metropolitan regions (Bay Area Economic Forum 2002). The five counties in the S.F. Peninsula and the East Bay Area represent the industrialized core of the region and have become dense and highly urbanized since the 1950s. In contrast, the four northern Counties remained more rural with significantly lower densities than in the core.

The region's population in the nine Bay Area counties had increased between 1980 und 2000 by almost 30 per cent, from about 5.2 million to about 6.7 million

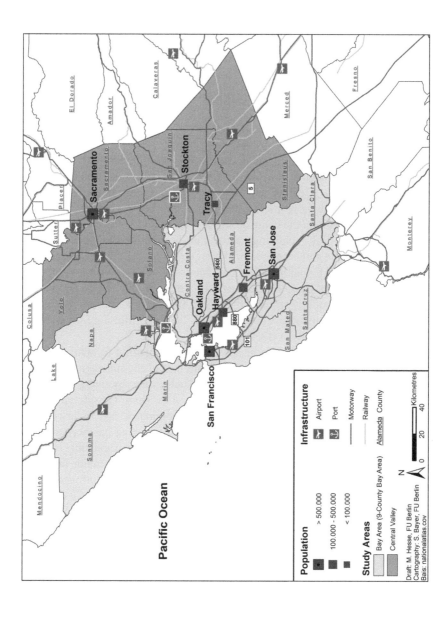

Map 5.1   The San Francisco Bay Area and the Central Valley

(ABAG 2003 based on U.S. decennial censuses). Most of this growth had occurred in San Jose in Santa Clara County (Silicon Valley) rather than in San Francisco or Oakland (see Table 5.1). The region's employment had increased from about 2.5 million in 1988 to about 3.4 million in 2001 (U.S. Census). Of this growth, a third had happened in Santa Clara County. In the beginning of the 2000s, economic development had been decreasing, for the first time in a decade, in the context of the decline of the New Economy (Bay Area Economic Forum 2002). Recently, a certain economic recovery has been observed, since the economic performance of the Bay Area had reached its "pre-bubble" state and composition in 2004 according to the latest Bay Area Economic Profile (Bay Area Economic Forum 2006).

The empirical investigation of this case study was carried out mainly in two areas: first in the East Bay area situated between the Cities of Richmond in the north and the City of Fremont in the south, which once was the region's main industrial corridor, providing the City of San Francisco with major industrial assets including the Port of Oakland and the Oakland International Airport. Today, this sub-region is both challenged by overall structural changes and the strong growth of the Silicon Valley. Light manufacturing with a high technology orientation, research and development, and professional businesses are moving north driving the much less competitive distribution and warehousing land uses out of the Bay Area and into the Central Valley. However, the intrusion of these freight-related land uses and transport businesses are being increasingly questioned and criticized by local residents, due to their impact on urban areas.

Correspondingly, as a regional reference point, a part of the Central Valley had been investigated, located about 100 kilometres north-west of the Bay Area.

Table 5.1   Population of the ten largest cities of the Bay Area, 1960–2000

	1960	1970	1980	1990	2000	*60–00*
San Jose	204.196	445.779	629.400	782.248	894.943	*338.3 %*
San Francisco	740.316	715.674	678.974	723.959	776.733	*4.9 %*
Oakland	367.548	361.561	339.337	372.242	399.484	*8.7 %*
Fremont	43.790	100.869	131.945	173.339	203.413	*364.5 %*
Santa Rosa	31.027	50.006	82.658	113.313	147.595	*375.7 %*
Hayward	72.700	93.058	93.585	111.498	140.030	*92.6 %*
Sunnyvale	52.898	95.408	106.618	117.229	131.760	*149.1 %*
Concord	36.000	85.164	103.763	111.348	121.780	*238.3 %*
Vallejo	60.877	66.733	80.303	109.199	116.760	*91.8 %*
Daly City	44.791	66.922	78.519	92.311	103.621	*131.3 %*

Source: U.S. Census 2003, own calculations

This part of Northern California has experienced strong growth since the early 1990s, which partly was a result of the dynamic development in the Bay Area and a resulting pressure on land use, transport etc. More recently the Central Valley is becoming a preferred location where corporations of the East Bay Area and the Silicon Valley were shifting to, not least those of the distribution business. Besides the different land rents, this has to do with good traffic access (regionally and california wide, with the Freeway I-5), lower labour costs and labour market supply. Due to such advantages, the Central Valley has changed from an agricultural seedbed to a distribution and light manufacturing oriented growth pole. It serves the Bay Area, the larger Northern and even Southern California regions. As a consequence of strong growth, this particular area has meanwhile been coined the "Third California" (Kotkin and Frey 2007), thus complementing the development of the Bay Area in Northern California and the Los Angeles-San Diego corridor in Southern California. The area investigated consists of Sacramento County (with the City of Sacramento as the California Capital) and Yolo County in its northern, and San Joaquin and Stanislaus County in its southern part, respectively. These Central Valley counties are characterized by significantly lower densities in terms of population and occupation compared with the Bay Area (except Sacramento). The sub-region appears as a prototypical case of de- and ex-urbanization which is initially derived from suburbanization in the Bay Area. Due to the increasing connectedness and interdependence of both parts, they are developing in a common socio-economic and spatial context.

Sub- and Ex-urbanization in the Bay Area

The historical roots of urbanization in the S.F. Bay Area date back to the mid-nineteenth century, the upcoming "gold-rush" and associated growth, first in San Francisco, afterwards on the eastern part of the Bay. Urban growth, significant immigration and the construction of the transcontinental railway line along the East Bay in 1869 were the initial impulses for spatial shift and dispersal. According to Walker (2001), the dispersal of industry is not to be understood as de-centralization, but as the cyclic regional concentration, first on the urban areas, afterwards on the southern Peninsula, with the establishment of new industrial districts and related suburbanization of working mens' districts. An accelerated de-concentration led to the establishment of refineries, steel mills, automobile factories, industry and chemical industrial plants. Due to specific product and process related cycles, the East Bay became the winner of these movements (Walker 2001, 44):

> All the same, the relative strength of labour and capital is insufficient to explain geographic change in a dynamic economy, without reference to the forces of technological and organizational change unleashed by industrialization. Fast-growing sectors can erupt in quite unexpected venues, while stagnant sectors and established sectors and established centres of industry fade away. The industrial base of California

has shifted repeatedly from era to era, recasting urban geography along the way. San Francisco embodied the eruptive stage in the mining era, the golden age of publishing, and early food processing. Thereafter, industries in which San Francisco had been a leader, such as mining machinery, men's clothing, beer and liquor, and leather, declined in significance in the state's economy. (Walker 2001, 44/45)

The next step in the process of (sub-) urbanization of the Bay Area was the dislocation of firms across the Bay towards Alameda County with Oakland becoming the new industrial centre, where commercial and industrial occupation was higher than in San Francisco already by 1910. Oakland profited much from the invention of the Central Pacific Railway from 1869, which helped establish an industrial corridor, including transport and logistics:

> With all sorts of goods passing through either coming or going, it was not long before factories multiplied in Oakland and in surrounding East Bay towns to process raw commodities and manufacture merchandise. Hides went to tanneries; lumber went to planning mills and carriage makers; hops and grain went to breweries; fruits and vegetables went to canneries. [...] Even if the raw product passed through San Francisco's port – salmon from Alaska, redwood from the north coast, sugar from Hawaii – much of it went to Oakland and vicinity for processing. (Bagwell 1982, 61f.)

In order to interpret the strong growth of Oakland (which belonged to the three fastest growing cities in the entire U.S. between 1900 and 1930), Walker did not emphasize the railway system, yet the structure and development of the industry: "Even in the twentieth century, the East Bay grew on water and rails, not trucks. But transportation was less significant than the restructuring of industry: the geographic shift to Oakland was driven above all by major re-orientations in sectoral composition and business organization in the region. The port and rail system grew to serve industry, not the other way round." (Walker 2001, 46) Sectors that were significantly growing were food processing in the then typical canneries, metal and machine tool industries, later on the automobile manufacturing and electronics. The more industrialized the East Bay was becoming, the more it would become spatially expanded:

> The new wave of industrialization stretched the metropolitan area of Alameda County dramatically north and east. Hand in hand with industry growth came extensive residential development and land speculation. As the westside and Emeryville built up their industrial base, the rest of the north county up through Berkeley and Albany filled in, creating a sea of small homes of the working class. During its period of growth from 1900 to 1930, the East Bay developed one of the most extensive streetcar systems in the country. The persistence of foot traffic makes it easy to connect the northward growth of the flatland to the suburbanization of industry, but trolleys and good wages allowed considerable lateral mobility; so workers" homes filled in the north-south core, hard against upper class redoubts in the foothills. [...] By the turn of the century, Oakland

was generating powerful burghers willing to battle with San Francisco over water supplies, port expansion, and industrial growth. (Walker 2001, 48f.)

Transport and distribution of resources, material and products were a major requirement for the rise of the East Bay to become the industrial backbone of the entire region. This effect was not only supported by the services of the railway and the port of Oakland, but also by the many transshipment stations operated by industry. The food processing industry not only developed the interfaces between waterway and railway, but also, by the end of the nineteenth century, the first large warehouses. Hence logistics were already an outcome of strong economic relationships between the agricultural parts of the region (e.g. in the southern East Bay) with the Central Valley. In the context of a particular division of labour, the financial capital for industrial expansion was provided from corporations in San Francisco, which altogether made the orchestration of the massive networks of transport and energy supply possible (cf. Walker 2001, 52). With accelerating technological progress, new means of transport and infrastructure were introduced and were shaping the region further, e.g. 1936 with the opening of the Bay Bridge and 1937 of the Golden Gate Bridge, as happened afterwards with the construction of the Freeway-network and also the BART-Metro. As a consequence, the industrial expansion of the East Bay continued. Before and particularly after the War, a significant growth of the region no longer originated from San Francisco, yet from places outside the Peninsula or even outside the Bay Area.

The Bay Area Becoming a Regional Metropolis

Beginning in the period after World War II, Scott (1959, 271) suggested the Bay Area becoming a "regional metropolis". In this context, industrial dispersal should play a major role again: places at the southern edge of the Peninsula and i.e. the East Bay performed highest growth rates. Again, this process did not follow the mere de-centralization of core city land use, yet the locational dynamics emanating from firms out of the region. Among these firms were now more and more high technology manufacturers, such as IBM. The still highly agricultural shape of the region was quickly changing, became industrialized and urbanized (see also Vance 1964).

> During all these years of economic peaks and setbacks the flood tides of suburban development kept rolling across the once rural landscape on the outskirts of the urban areas. [...] One urban wave swept southward along the bay side of the San Francisco Peninsula into northern Santa Clara County, creating an almost unbroken pattern of low-density development. Other waves surged outward in all directions from the San Jose area. [...] On the eastern side of the bay another flood of development advanced southward past Hayward into the green acres of Washington Township. And in the San Ramon and Ygnacio valleys of Contra Costa County, east of the Berkeley Hills, still other waves of development deplaced walnut groves, apricot orchards, and vineyards.

The northern counties experienced to a lesser extent the same kind of "runaway" suburbanization. (Scott 1959, 280)

Urban expansion was now increasingly driven by infrastructure investments, i.e. the freeway system. Accessibility became improved, and settlements could further expand. Even today, the major pattern of urbanization is based on the freeways network with the I-110 along the Peninsula and the I-80/880 in the East Bay. From the 1950s and 1960s on, the South Bay Area, i.e. Santa Clara County were in the focus of industrial development, based on entrepreneurship, defense spending and regional networks – as a successful combination of "competition and community" (Saxenian 1994). During the 1990s, the advent of the Internet created another boost that had been disrupted only temporarily. In terms of settlement, the bigger inner cities had already been completed by the turn of the 19th and 20th Centuries. As part of the next wave of urbanization, the inner suburbs had been added. Since the 1940s, with increasing private motorization, the first modern suburbs emerged, ring-wise around the medium and smaller sized cities. A third spatial shift has driven new developments into outer areas even beyond the region since the 1980s. Now the Bay Area faced increasing suburbanization at its periphery, even in non-integrated, peri-urban areas (vgl. ABAG 1998, Vance 1964). Lang (2003, 72) called the San Francisco Bay Area a "fragmented metropolis", that includes, besides the three downtowns San Francisco, Oakland and San Jose, also four edge cities (North San Jose, Pleasanton, San Ramon und Walnut Creek) – and also numerous "edgeless cities", settlements without a clear structure, being neither centre nor city on the edge.

As the main indicator of the emergence of edgeless cities, Lang uses the spatial distribution of employment in the service sectors. Indeed, the differentiated development of occupation is strongly associated with the spatial shift from the core city to suburbs and beyond. Landis and Reilly (2003) have distinguished three major waves of regional development in California. The year 1950 is considered the turning point in the relationship between core cities and suburbs: from this point on, employment growth was higher and faster in the suburban areas than it was in the core cities. Since 1980, additional employment had almost exclusively occurred in the suburbs. In the 1990s, the Bay Area had experienced an employment growth of about 13.5 per cent not at least in the context of the New Economy and the Silicon Valley. In spatial terms, this growth had ocurred in a quite differentiated manner: whereas the sector within ten kilometres from the old centres had been growing, yet to an disproportionately low extent, growth rates were higher with increasing distance from the core. The highest growth rates were observed in the zone that was 40 or even 50 kilometres away from the urbanized core (see Table 5.2). In absolute numbers the zone in a distance of 20 and 30 kilometres from the centre performed best (ebd.). As a consequence, the Central Valley is developing to a complementary space for the Bay Area – the higher the pressure of land rents and more competitive land use is becoming.

Table 5.2   Spatial Distribution of Bay Area Employment, 1990–2000

	10–km Ring	Employees		
		1990	2000	1990–2000 in %
	1–10	1,992,600	2,180,213	9.4 %
	11–20	490,930	560,260	14.1 %
San Francisco Bay Area (9–County)	21–30	376,780	441,687	17.2 %
	31–40	216,110	271,233	25.5 %
	41–50	17,770	22,537	26.8 %
	Total	3,094,190	3,512,717	13.5 %

Source: own after Landis/Reilly 2003, 21

Sub- or Ex-urbanization into the Central Valley

De-centralization of settlements is expected to continue in the near future. In this respect, the areas adjacent to the Bay Area are considered as the most dynamic ones. This also applies to the four Central Valley-Counties Sacramento, San Joaquin, Stanislaus and Yolo that were investigated in the context of this case study. The San Joaquin Valley stretches along the San Joaquin River and is bounded by the cities of Stockton in the north and Bakersfield in the south representing a corridor of about 350 kilometres. It was once the traditional central agricultural area of California and only sparsely populated until the mid-twentieth century. The construction of the Highway 99 fostered urbanization along the corridor, which was initially confined to Greater Sacramento. The expansion into the San Joaquin Valley that happened afterwards was supported by the construction of the Interstate I-5 that connects the northern and southern parts of California. With the amazing growth rates of population and economic development in the Bay Area of the 1980s and 1990s, the San Joaquin Valley became attractive not longer as an agricultural production space only, yet increasingly as a complementary housing and commercial area. Population forecasts do confirm that such a growth perspective is likely to take place in the coming future as well (see Table 5.3).

Land in cities such as Tracy and Stockton, Manteca, Modesto or even Sacramento could easily compete with the high land rents of the Bay Area (see Figure 5.1). Although the Central Valley belonged to the leading areas of the global agribusiness, an increasing amount of agricultural land is being transformed and developed. Since the 1990s, commuter corridors have emerged between Stockton, Modesto, Tracy and the Bay Area, using the Freeway (I-5, I-580) and also Amtrak rail services. As it is often the case in advanced suburbanization processes, particularly the smallest municipalities are likely to perform highest growth rates. This is expected to continue in the near future which contributes to

Table 5.3   Population forecast for Central Valley counties until 2040

	Population 2000 and Forecast 2020/2040*			2000–2040 in %
	2000	2020	2040	2000–2020
Sacramento	1,212,527	1,651,765	2,122,769	75.1 %
San Joaquin	579,172	884,375	1,250,610	115.6 %
Stanislaus	459,025	708,950	998,906	117.9 %
Yolo	164,010	225,321	298,350	81.9 %
Total	*2,267,125*	*2,832,356*	*4,670,635*	*93.4 %*

Source: Landis/Reilly 2003/The Great Valley Center 2003

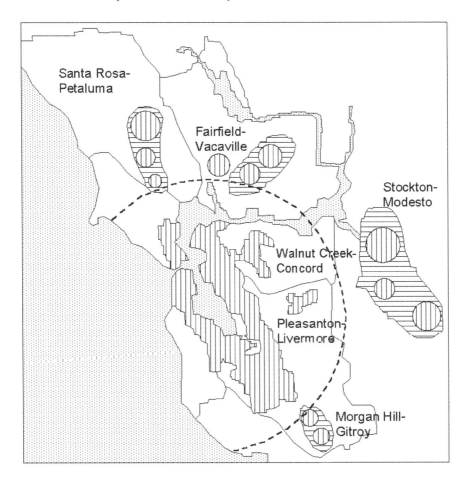

Figure 5.1   Sub- and Ex-urbanization in the Bay Area and beyond
Source: Own after Race (2001)

the forced urbanization of the Central Valley. Other things being equal, a possible consequence within just one or two decades could be the emergence of a core, ribbon-like city region across the Central Valley stretching between the City of Redding in the north and the City of Bakersfield in the south. The main question of this case study is to what extent logistics and freight distribution might contribute to such a development, or whether firms may prefer staying closer to the industrial hot spots in the Bay Area.

Locational Dynamics of Logistics and Freight Distribution

The Region as a Distribution Area

Among the metropolitan regions of North America, the San Francisco Bay Area does not belong to the major hubs of the inter-regional goods exchange as it applies to Los Angeles/Long Beach, Seattle/Tacoma or New York/New Jersey. However, even if the Bay Area does not concentrate global goods flows at a large scale, it represents without doubt an important node within the national and also international network of logistics. The region provides infrastructure of all major transport modes, which consists of trucking, rail freight, ocean and inland shipping, also air freight. Judging from the logistical standpoint, the Bay Area represents a major production space particularly for corporations in the Silicon Valley. This property makes the region develop a positive export-import ratio meaning that more goods are being shipped away than are being received from other destinations. Also, the region represents an important consumption space, with about 7 million people who need to be provided with goods. Also different from the other global hubs in North America named above, the structure of goods traffic of the Bay Area is disproportionately highly regionally based: 46 per cent of the goods tonnage that is transshipped in the Bay Area remains within the region or within California. There is also a lot of exchange going on with the rest of California and Southern California (Los Angeles) respectively but also to a significant degree with the Central Valley (Cambridge Systematics et al. 2003).

The most intensive goods exchange is being practised between Alameda and Santa Clara Counties, also between Alameda and Contra Costa Counties. The central role of Alameda is also related to the two big interfaces of the international goods flow located in the County: first the Port of Oakland, and second Oakland International Airport. These two facilities represent the major gateways for incoming and outgoing regional freight flows. At the Port of Oakland, the two big remaining railway companies of the American West, Union Pacific (UP) and Burlington Northern Santa Fe (BNSF) operate an intermodal transport terminal. BNSF also operates a terminal in Richmond and in the North Bay (dedicated to UPS). UP maintains a hinterland connection via Roseville, BNSF does via Stockton into the Central Valley.

The largest portion of foreign trade of regional firms is being operated in relation with Asia; the most important logistics interface of the region, the Port of Oakland, operates most of its transport relations with Asian destinations (China, Japan, Taiwan, Korea, Hong Kong, Thailand) and also New Zealand, whereas highest growth rates have been observed regarding the relationships with NAFTA-countries Canada and Mexico. Among the exported goods, computers and high tech-equipment and also machinery comprise the most important segment (37 per cent of consignments, based on the year 2005), followed by electronic equipment (35 per cent), measurement and control devices with 9 per cent and also vehicles (4 per cent). During the 1990s, the value of exports only out of the three metropolises San Jose, San Francisco and Oakland grew by 49.5 per cent. San Jose and the Silicon Valley, respectively counted to the biggest share of this growth.

Trucking is by far the most important freight transport mode in the Bay Area, with a share of about 80 per cent of the tonnage and almost 82 per cent of the goods value. The freight railway has a market share of about 6 per cent of the tonnage and 12.5 per cent of the value, shipping accounts for about 13 per cent and 3.5 per cent respectively. During the 1990s, the volume of transshipments handled in the Port of Oakland, which is the 4th largest container port of the nation (based on TEUs), has increased by an annual rate of 5 per cent. Such growth is also being expected to occur in the upcoming two decades. Airfreight, despite being low in terms of its share of the entire goods transport market, is supposed to grow at the fastest pace even in the near future: its tonnage is supposed to triple over the next 20 to 30 years (ibid.).

The rising volume of foreign trade and the economic growth in general have triggered a significant increase of transport flows as well. This exerts a high pressure on transport infrastructure, since existing capacity cannot be extended accordingly. Any expansion is critical due to the high densities in the Bay Area, and also as a consequence of the complicated political setting in an highly urbanized region. The Port of Oakland is a good example in this respect, since there is an urgent need to enlarge its terminal areas and infrastructures for adding to goods handling capacity, in order to compete with the other big ports at the West Coast (Los Angeles/Long Beach and Seattle/Tacoma). Due to the inner-city location of the port and its disturbing influence on the adjacent neighbourhoods, many conflicts need to be solved. The three big airports of the Bay Area, San Francisco International (SFO), San Jose International and Oakland International (OAK), are challenged to further extend their runways. Oakland International already requires additional space for the handling of air-freight, and a new runway is predicted to be necessary by 2008.

Road freight transport capacity is also scarce on most of the important sections of the freeway-network of the region. Given the economic significance of timely delivery, this causes serious problems for the region. Particularly the major arterials of the Bay Area, the I-80/880 corridor along the East Bay, the I-580 into the Central Valley, the I-680 from Contra Costa County into the Silicon Valley and the U.S. 101 along the S.F.-Peninsula are almost permanently congested. Trucking normally contributes to a share of 6 up to 8 per cent of the average daily highway traffic; with these values climbing up to 20 per cent on the Central Valley I-580 and I-5 highways.

According to the increasing spatial shift out of the Bay Area into the Central Valley, the Bay counties have developed strong relationships to places in the Valley. One reason for this is the historical fact that many of the agricultural commodities produced in the Valley were traditionally distributed via the interfaces in the Bay Area (i.e. ports and airports). Also, if the Central Valley is increasingly developing as a location complementary to, or corresponding with the Bay Area, this is likely to trigger a rising number of freight and passenger flows between the two subareas. This is even more the case once many firms in the Bay Area depend on acquiring a medium or low paid workforce. Nonetheless, their salaries usually do not cover the high cost of living in the Bay Area. The more households and firms are seeking for affordable, cheaper locations in the Valley, the Interstate I-580 will function as a gateway for flows from the Bay into the Central Valley and vice versa; keeping in mind the Interstate I-5 in the core area of the Valley, this might be the case in the rest of California as well. Not coincidentally, the freeway segment between both sub-regions accounts for the most congested in the entire Bay Area.

The Spatial Organization of Warehousing and Logistics

The spatial dynamics of goods distribution facilities and flows in the East Bay Area and the Central Valley have led to three different types of location, with separate development patterns and different potentials for the future. First, there are main hubs, such as the Port of Oakland and Oakland International Airport, which have undergone strong growth in the past, especially over the last 12 years. This also applies to the San Francisco International Airport on the Peninsula, which serves parts of the East Bay as well. Secondly, there are traditional, more spacious and less nodal goods distribution areas in the East Bay – that is, in the Cities of Oakland (e.g. West-Oakland), San Leandro, and Hayward. These places are partly related to the main hubs, but they are locations that also contain businesses which are not or no longer related to goods distribution. In the East Bay area, these three municipalities contain more than 53 million square feet of warehousing space, estimated to be almost 60 per cent of the building base of the entire sub-region (BT Commercial Real Estate 2001). Currently, the three cities are maintaining their high amount of warehousing space (building base), and have some minor additions, but most of the growth is taking place elsewhere. According to the county and city employment data, the growth rates in Standard Industrial Codes (SIC) for the distribution industry (4200, 4400, 4500, 50–5200) are the highest in the outer areas of the Bay (except for Alameda County, with Oakland Port and Airport) and the northern Central Valley. Third, many of the new warehousing and DC sites have been established in existing commercial areas of the Central Valley, which is now supplying the Bay Area with goods. In two of the most dynamically growing locations, Stockton and Tracy, goods distribution firms account for about 80 per cent of all firms in the new commercial areas. Interviews with real estate personnel revealed that they estimate that 90 per cent of these firms have moved there from the Bay Area.

Since transport infrastructure certainly is a vital component of the locational dynamics of distribution firms, a large number of distribution and logistics firms are located in the East Bay along the I-80/I-880. Within the East Bay Area, freight transport and handling activities are traditionally concentrated in Alameda County, especially in and around the Cities of Oakland and Hayward. Although this is a dense, highly populated and, judging from land rents, expensive area, the distribution sector will play a major role in this particular area in the future as well, since these places offer important locational advantages. They are not only close to the main hubs mentioned above, yet, they connect the East Bay with the S.F. Peninsula by Bay crossing bridges. This is particularly true for the Hayward area. However, such original locational advantages are being challenged by effects that are based on push-forces emanating in the Southern Bay Area, i.e. Santa Clara County. Transport and distribution firms are usually too weak in terms of rent and capital base to successfully compete with firms from the high tech- and Service sector.

The local submarkets for commercial real estate deploy an immediate response to these changing land use relationships and competition. Insofar, they document the degree of regional economic change properly. Table 5.4 reveals the stock on warehousing space that is available on the market (which means: currently not used), also the rents (in US-Dollar per square foot). The data base is delivered by a commercial real estate firm and is considered a useful source for assessing the building block and transactions of warehousing space in each subarea; comparative data provided by other firms support these findings. The data gives evidence first to the significance of the East Bay as a prime location for warehousing, logistics and distribution, concentrating as much as half of the entire warehousing space of the entire Bay Area. These places are characterized by the lowest rents paid for warehousing space; the average rent level accounts for almost 50 per cent of that in San Mateo County and even less than the respective numbers of Santa Clara County. In contrast, vacancy was periodically the highest in Santa Clara County, which was due to the ups and, particularly, downs of the New Economy. The decline of high-tech manufacturing and marketing activity has also led to a decreasing demand for warehousing space. As a consequence, firms that supply warehousing space had shifted their facilities toward alternative locations, in order to avoid the high rents in the south Bay Area. One response of the high tech-firms in this respect is to make use of the so-called flex-buildings, which are hybrids of warehousing, service and production spaces. These mixes may make it difficult to relate certain buildings and areas to their main use category, so the real estate-statistics do not always mirror the related land use correctly. In general, the main assumptions made on the spatial distribution of logistics related land use remain unquestioned by this detail.

**Table 5.4   Major Bay Area markets for warehouse space[a]
(first quarter 2003)**

Area	Total Building Space (Sq. Ft.)[b]	Per cent of Total Space
Mission / SOMA	5,650,538	
3rd St. Corridor / Potrero Hill	10,534,099	
Bayview	4,590,940	
Total San Francisco	20,775,577	11%
Brisbane	4,336,936	
South San Francisco / San Bruno	18,780,077	
Burlingame / Millbrae	3,775,140	
San Mateo / Foster City	720,612	
Belmont / San Carlos	3,398,118	
Redwood City	1,033,304	
Menlo Park	1,861,136	
Total San Mateo County	33,905,323	19%
Sunnyvale	3,275,858	
Santa Clara	3,912,332	
North San Josè	9,933,635	
South / Central San Josè	8,876,284	
Morgan Hill / Gilroy	2,319,580	
Milpitas	6,606,273	
Total Santa Clara County	34,923,962	19%
Richmond	4,746,259	
Berkeley	2,085,950	
Emeryville	2,111,411	
Oakland	15,646,379	
San Leandro	16,193,677	
Hayward	20,159,811	
Union City	8,333,335	
Newark	3,714,043	
Fremount	8,598,612	
Total East Bay I-80/880 Corridor	81,589,477	45%
Livermore	6,459,694	
Pleasanton	2,614,017	
Dublin / San Ramon	2,627,059	
Total East Bay Tri-Valley	11,700,770	6%
Grand Total	182,895,109	100%

Note: The data for all market areas except Pleasanton and Dublin / San Ramon in the Tri-Valley area are from Research Reports prepared by BT Commercial Real Estate, for First Quarter 2003. Date for the two Tri-Valley areas are from CB Richard Ellis Industrial Market Reports for First Quarter 2003.
[a] The warehouse market includes buildings typically used for bulk warehouse purposes with clear heights of 18 feet or more, dock and/or loading grade doors, minimal build-out, and limited glass.
[b] Total building space includes space in warehouse buildings over 10,000 square feet in size.

Sources: BT Commercial Real Estate; CB Richard Ellis; Hausrath Economics Group.

Table 5.5   Changes in industrial rents and vacancy rates for Major Bay Area markets, 1995–2002/03

	1995	1997	1998	2000	2002	2003	Change 1995–2002
Rents ($ per sq. ft. per mo.)							
Warehouse	0.37	0.47	0.54	0.72	0.52	0.50	+41%
Manufacturing	0.41	0.57	0.77	1.30	0.67	0.62	+66%
R&D	0.79	1.51	1.58	4.00	1.29	1.20	+63%
Vacancy Rates							
Warehouse	6.8%	4.9%	3.9%	2.8%	9.1%	10.6%	
Manufacturing	5.9%	4.1%	3.1%	2.1%	6.7%	7.3%	
R&D	7.2%	4.8%	9.5%	3.4%	20.5%	21.7%	

Note: Major Bay Area industrial markets include San Francisco, San Mateo County, Santa Clara County, and East Bay I-80/880 corridor, as identified in Tables 2,3, and 4. Comparable trend date is not available for more outlying industrial markets.

Within the East Bay Area, which can be considered the logistical backbone of the region, almost 60 per cent of the warehousing space is concentrated in the three subareas Oakland, San Leandro and Hayward. The City of Hayward, with a population of 140,000, has almost as much warehousing space as the City of San Francisco has, which is five times bigger judging from its population. A significant supply of warehousing space in the East Bay ensures constantly low rates to be paid for, and, as long as the traffic conditions allow for, these are optimal locations for the provision of firms in the South Bay Area, where warehousing space is expensive or extremely scarce, and new supply remains limited.

The land rent differential that is included here basically reflects the model of centre and periphery (see Table 5.5). However, this picture is highly differentiated by the poly-centric structure of the Bay Area. The differences between the subareas are significant, and they are even higher, the more distant parts of the Bay Area's hinterland are taken into account: In Benicia, at the northern edge of the East Bay, warehousing rents account for just 64 per cent of the average level that is found in Oakland. In the Central Valley, these numbers drop to a level which is half of the average rents of the East Bay (Tioga Group et al. 2001, 14). The more the traffic on the central corridors of the East Bay is becoming congested, the higher is the incentive (or pressure) for firms to move their facilities to places beyond the Bay Area.

Logistics Employment

In order to track the spatial distribution of logistics in more detail, employment data on relevant sectors have been assessed, particularly the *County Business Patterns* issued by the U.S. Census Bureau were used as an indicator. Thus a logistics and freight distribution segment had been compounded on the basis of the respective SIC-codes (*Standardized Industrial Classification*). In comparison with the Berlin-Brandenburg data, this segment comprised Trucking and Warehousing (SIC 4200), Freight transport arrangement (SIC 4730), Marine Cargo handling (SIC 4491), and separated from logistics also Wholesale trade (SIC 50). For the period 1998 ff., the new North American classification NAICS has been used. The related subsections were Trucking (NAICS 484), Freight transportation arrangement (NAICS 4885), Couriers and Messengers (NAICS 492), and Warehousing and Storage (NAICS 493), in a separate segment also Wholesale trade (NAICS 42). In order to establish a time series, the data comprising 1988 to 1997, and 1998 to 2004 respectively, had been compared, since they are based on identical classifications. With some minor changes in classification in mind, a time series for the period 1988 to 2004 was also possible, yet may deliver only limited evidence due to the change from SIC to NAICS.

Logistics related employment in the nine counties of the Bay Area has been decreasing between 1998 and 2004 by about 22 per cent (see Table 5.6). This decrease is significantly lower than the growth of this sector had been in the entire U.S., which achieved 27 per cent; however, it was even much higher than compared to the average growth of total employment in the same period (about 7 per cent). Among the counties with the most intensive decrease in absolute and relative numbers in the Bay Area were Alameda County and Contra Costa County – the most centrally located and highly industrialized areas of the East Bay, also

Table 5.6   **Employment in selected logistics groups in the Counties of the Bay Area and the Central Valley 1998–2004, also in the U.S.**

	1988	1996	1998	2001	2004	*98–04*
Bay Area	73,396	81,493	57,707	54,572	44,646	*-21,8*
Central Valley	16,171	18,425	23,426	25,733	28,650	*22,3*
Summe	89,567	99,915	81,133	80,305	73,296	*-9,6*
USA	1,647,826	2,040,077	2,134,523	2,288,865	2,728,324	*27,8*

Source: U.S. Census Bureau, County Business Patterns; own calculations

1988/1996: SIC (4200: Trucking; 4400: Warehousing/Storage; not included 5000: Wholesale trade)
1998–2004: NAICS (484: Trucking; 4885: Freight forwarding; 492: Courier, Express Services and Messengers; 493: Warehousing/Storage; without 42: Wholesale trade).

San Francisco County as a highly urbanized, inner-city area. The respective decline of employment corresponds with long-term trends in urban land use changes. However, logistics employment in San Mateo County (where the San Francisco International Airport is located) and Santa Clara County, including Silicon Valley, has increased recently, whereas it had been declining between 1998 and 2001. The rural counties in the Northern Bay Area (Napa, Sonoma, Solano) have developed in a positive way, though they were starting from a very low basis. The accelerated decline of employment related to logistics in Alameda County is somehow surprising, since Alameda is one of its logistics centres and still hosts the intermodal hubs of the Bay Area. Alameda has lost employment in those segments that had been strongly developing earlier, such as courier and express. Among the logistics subsectors, the large group of couriers and messengers contributed most to the decline of logistics employment in Alameda County. The logistics percentage in Alameda County dropped from 2.75 per cent in 1998 to 1.63 per cent in 2004.

In contrast, the Central Valley had in fact been developing to a new industrial space over the last two decades, thus complementing the supply of the dense core areas of the Bay that are increasingly facing the disadvantage of agglomeration. The subareas of the Central Valley had achieved growth rates in logistics employment between 1998 and 2004 that almost exactly equals the respective decline in the Bay Area. The enormous growth of the Central Valley is particularly driven by the strong performance of San Joaquin County, which added more than 57 per cent during the period covered by the data, even from a relatively high starting level. The growth rates were highest with regard to couriers and messengers and to warehousing and storage – exactly those segments that were losing in the East Bay – for a considerable amount of time. The logistics percentage of San Joaquin County increased from almost 5 per cent in 1998 to more than 6.5 per cent in 2004. The shift of existing facilities towards the Central Valley or the establishment of new sites seems to be a logical consequence of the changing locational conditions. However, the counties in the Central Valley did not totally improve in terms of logistics employment, as the loss of Sacramento County reveals.

In general, it also applies to the areas with increasing employment that the logistics growth rates tend to exceed the growth of total employment significantly. In comparison, it also seems to be evident that all Central Valley-counties rank specifically higher than the Bay Area-counties do. Except one county (Sacramento County), the Central Valley performed well since 1988, whereas the numbers in the Bay Area are decreasing since the mid-1990s, a tendency that became even accelerated since 2001. It is also remarkable that, if one adds the logistics employment for both sub-regions in the year 2004, the result is almost 10 per cent below the level of 1998. This means that, despite sectoral and spatial shifts and also regardless regionally differentiated growth, there is rationalization at work that does not allow assuming an overall growth would take place in current logistics related employment.

Different from the southern Bay Area (Santa Clara County), where almost no new warehouses had been built over the last years, there is still a relevant demand for warehousing space and distribution centres in the East Bay Area. In this part of the region, a certain supply is still provided. These are partly new developments, mostly subject to the re-use and re-development of existing space. Firms that are selecting these spaces are not only challenged by rising land rents and increasing locational competition caused by service and high-tech firms that can afford to pay much higher rents (see above). They also face a critical assessment of transport related activities enforced by municipalities. Particularly environmental and community issues are being addressed, with respect to air pollution and noise emissions, land use issues and alike. One point that is oftenly made is how low income neighbourhoods, that are often adjacent to port areas and distribution centres, can be better protected; another is related to the possibilities of offering access to labour markets, particularly for low-paid jobs.

> Most importantly, the Third California remains perhaps the greatest untapped outlet for upward mobility in the Golden State. In some senses, this reflects as well the difficulty of wealthier areas—such as First California's San Francisco Bay Area and, to a lesser extent, coastal Second California in the south—to provide new jobs and opportunities, especially opportunities for homeownership. (Kotkin and Frey 2007, 2)

However, as the numbers above indicate, logistics does not represent a "boom"-sector per se. So there is no reason to follow the simple assumption, logistics would compensate for the job-losses caused by de-industrialization in old-industrialized regions. According to the findings of this case study, economic development managers of all municipalities and the two counties in the East Bay primarily aim at attracting investments of high-tech and service-oriented firms. Transport and distribution firms are no longer on their agenda, since they are critical in terms of emissions and disturbances, whereas their contribution to tax spending remains limited. So, by adding this to the dynamics of the land market and the pressure of congestion, it is also an active political regulation of municipalities and counties that pushes these firms out of the agglomeration.

The respective complementary area for the East Bay is about to emerge in the Central Valley. For decades, the Valley had been serving the agglomerations in the Bay Area and in Southern California (Los Angeles, San Diego) with water and agricultural supply (see above). More recently, this functional space has been changing rapidly, particularly under the influence of urbanization in the Bay Area, yet also in the context of an endogenous path of industrial development. The region's response to the disadvantages of agglomeration deployed in the Bay Area consists of cheap land, freeway-access and a broad labour pool that originally stems from the agricultural basis of the area. These locational settings are extremely valuable from the distribution businesses' perspective. Hence it comes without surprise that the Central Valley sub-regions have achieved highest growth rates in the logistics employment over the last 15 years. In this regard, the strong growth of San Joaquin County seems to be most significant, with the trucking and warehousing and storage sections developing at fastest pace.

The related growth of logistics employment in the Central Valley-counties has been continuously higher than the growth of total employment was, that is what the location quotient of the Central Valley reveals (see Table 5.7 for selected Central Valley counties). Compared with the Bay Area, it is also remarkable that all Central Valley counties are ranking higher than the Bay Area-Counties do, except Alameda County. In general, the development of logistics employment numbers in the Central Valley turns out to be positive since 1988, whereas the related development in the Bay Area follows just the opposite direction.

The spatial variation of logistics and freight distribution firms shall be assessed in the context of different patterns of urbanization. Some of the logistical core areas still perform strongly, and they are closely connected with the functional centres. However, each of these centres is bringing about specific peripheries which contribute to establishing the polycentric region of the Bay Area as a whole. This type of region can no longer be analysed according to the traditional dichotomy of centre and periphery: different from the Berlin-Brandenburg region, the Bay Area appears multifaceted, heterogeneously structured with complex patterns of networks across the traditional hierarchy of centres. This leads to two consequences: first, the focus is shifting from centre-periphery relationships to different partitions of suburbanization, particularly at those places that are located more remote from the old core. Second, the particular contribution of logistics to suburbanization appears difficult to identify, since the urbanized landscape is much more heterogeneous than it was before. However, the de-concentration of logistics and distribution land uses appears to be evident.

Table 5.7   Location quotient of logistics employment in selected counties of Northern California

	1998	2004
Alameda	1.59	1.17
Contra Costa	0.46	0.51
San Mateo	1.12	1.05
San Francisco	0.83	0.27
Sacramento	1.31	0.68
Santa Clara	0.39	0.36
San Joaquin	2.72	2.76
Stanislaus	0.96	0.91
Yolo	2.53	4.90

Source: U.S. Census Bureau, own calculations

NAICS (484; 4885; 492; 493).

The spatial shift of logistics and distribution firms towards the Central Valley can be therefore interpreted as a significant contribution of this sector to suburbanization processes, as these developments are heading towards more distant areas away from the older core, yet remain strongly connected to the agglomeration. This will be exemplified in the next section of this chapter, where selected commercial and industrial areas are being presented. The related pattern of sub-urbanization also tends to be an urbanization of the adjacent regions. Insofar, a certain ex-urbanization of distribution, in a sense that there is a movement of firms underway out of the agglomeration towards peripheral regions of the Central Valley, is not yet at work in this particular case. The attraction of distribution investments occurs in an area that is currently under the influence of accelerated urbanization. Tendencies of ex-urbanization may be the case in California once developments take place in much more distant areas, e.g. in low-tax areas of the State of Nevada. According to observers, a significant amount of firms has shifted over there in order to improve their cost basis. This is also true for firms in the logistics and distribution business. In contrast to those ex- or peri-urban places, the counties of the Central Valley are particularly attractive for those distribution businesses which have their major customers located in the core Bay Area. As a consequence, and according to future scenarios for the next two decades, disproportionately high growth rates are being expected for San Joaquin County (Cambridge Systematics et al. 2003, Hausrath Economics et al. 2003).

Spaces of Distribution: Contrasting the East Bay and the Central Valley

In this subsection, two different areas that are extremely important for goods movement and logistics in the Northern California region are being presented. These cases unfold the locational dynamics and framework conditions in a broader context, to which each single firm tends to respond individually.

a) Core city logistics centre: The Port of Oakland, West/East Oakland   The Port of Oakland and the related infrastructures are the main attractors of a significant concentration of distribution activity in the East Bay, particularly in Alameda County, alongside with extremely good accessibility of major parts of the entire region. The historical background for this kind of agglomeration are the traditional linkages between the on-site commodity handling processes and the related services, such as storage, packaging, pre-processing, financing, labour supply etc. These linkages have vanished, due to the integrated management of logistics operations, and are now being organized at different segments of the chain, either on the ship, at the port or in the hinterland, respectively. Containerization remains the main attractor of business operations at the port or around it. Consequently, customs, import- and export firms, packaging and pre-production or distribution are functions that still favour port proximity and port related locations: some of them are even port-dependent. This particular industry mix predominates the commercial areas adjacent to the Port of Oakland. The Port is located close to the

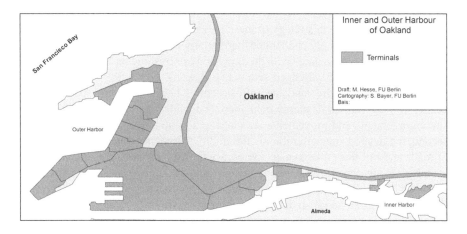

Map 5.2   The Port of Oakland

Source: Own

entrance of the Bay west of the Bay Bridge and east of downtown Oakland (see Map 5.2). This central urban location offers many advantages on one hand. However, the historical proximity to inner-city districts also causes serious conflicts on the other hand. This applies particularly to the West-Oakland neighbourhood that hosts a low-income and deprived community that is exposed to traffic noise and diesel-engine emissions, due to the extraordinary amount of port related trucking activity that takes place in an area much larger than the port is.

Today, the Port of Oakland is challenged particularly by competition with the ports of Los Angeles, Long Beach and Seattle at the North American Westcoast. Oakland has been able to increase its volume of container transshipments, particularly due to the strong output of the high-tech industries in the Silicon Valley since the 1980s and, more recently, in the context of rising trade relations with the emerging Asian-Pacific economies. The transition from a main industry- and bulk-related port to a major container port over the last two or three decades is considered successfully managed. As a result, future predictions are expecting a further expansion with even more growth in terms of transshipment.

According to the widely practised economies of scale in maritime shipping, the Port of Oakland aims at further extending and mechanizing terminal operations, also at expanding its land base, in order to prepare for the future rise of container handlings (Port of Oakland 2000). Therefore, the port-side infrastructures are being expanded and two new container-terminals have been constructed on a former U.S.-Navy site. Also, a joint intermodal-terminal for rail-barge transshipment has been built. The dredge of the S.F. Bay has been deepened up to 50 feet in order adapt to the requirements of "Post-Panamax"-container vessels. The availability of additional land from the adjacent Oakland Army Base will now enable the Port to significantly expand the existing Outer Harbour Terminals, and to replace them with modern terminal technology.

However, any growth strategy of these mega-hubs within the system of global flows appears increasingly challenged, particularly due to three factors: first, there are enormous costs associated with dredging and establishing underwater channels (and the disposal of the dredged sediments as well), also with the construction and maintenance of the landside infrastructure. Even the mere space that is needed to park trailers or empty containers is actually scarce (as the related on-site arrangements and movements are causing additional demand for transport); second, the desired concentration of commodity shipments causes the congestion of roads and rail lines that connect the port with its hinterland. This congestion is even increasing, the more successful the port may become in receiving calls from container lines and increases its throughput; third, there is a rising conflict regarding the waterfront from an urban development perspective, since redevelopment plans are always being discussed as a potential alternative to the costly strategies of port operation and expansion. At least the two challenges mentioned first also apply to the Port of Oakland which not only needs to expand its market share, yet is also committed to improve both accessibility and functionality of its operation and the acceptability of the related community impact.

In order to cope with these two challenges, the port needs to become more efficient in terms of logistics and distribution operations, and it is also committed to make a better use of land resources, being this on-site of the port or in association with more distant, satellite-like locations. Regarding the first issue, the traffic conditions on the major corridors serving the port (I-80, I-580, I-880) are already critical and expected to become even worse in the future. Passenger travel on the corridors is likely to increase as well. Other conditions being equal, there will be the urgent need for improving the land-side connections of the port. An intermodal transport-bridge into and out of the port is discussed as one of the desired solutions. Consequently, in October 2004, the Port of Oakland announced a new alliance formed between the Port, Northwest Container Services, Inc. and the City of Shafter (CA), which probably will result in an improvement of goods movement in Northern California. Northwest Container Services will be serving an integrated logistics centre which is located in (and developed by) the City of Shafter. The centre is supposed to combine an inland intermodal facility with dedicated rail logistics serving international marine terminals at the Port of Oakland. This inland intermodal centre is dedicated to improve the hinterland links into and out of the Port of Oakland and thus to expand the handling capacity of the port significantly. It includes a rail freight intermodal facility that will support the re-use of inbound (import) containers, loaded with consumer goods destined for the Central Valley and Southern California, as outbound containers for export of agricultural goods from the U.S. For almost two decades, NWCS has been operating the shuttle train service between Seattle, Tacoma and Portland in the Pacific Northwest. The rail shuttle is almost similar to what had already been proposed as the "California Inter-Regional Intermodal System (CIRIS)" in the Port Location Study (Tioga Group et al. 2001). In addition, since 2005 the MOL container line is offering a daily express service to ports in South China (including Hong Kong) which is further connected by the Burlington Northern Santa Fe Railroad (BNSF) to the North American Mid-west. BNSF operates a rail facility on the site of the Port of Oakland.

To improve the land related situation at the Port of Oakland, two studies had been conducted recently, the Port Location Study (The Tioga Group 2001) and the land use related reports in the context of the Bay Area Goods Movement Study (Hausrath Economics 2003, 2004). Both studies tied up to the problems described above, analysed the development of the freight and logistics related real estate market and were seeking for land use related solutions of the problem. These problems are not only important for the Port of Oakland, yet also for the majority of the logistics and freight distribution firms which are located in the core East Bay Area and in the vicinity of the port. The industrial sites that are being used by these firms are highly valued, because of short distances to the port facilities and the possibility of efficient vehicle operations, container and trailer parking etc. close to port and company location. However, given the general traffic increase and the expected rise in the volume of containers, the sites that are mainly located in the cities of Oakland, San Leandro and Hayward appear to be increasingly problematic, both in terms of the traffic conditions along the I-80/880 corridor and of the acceptability with respect to community issues.

The aim of the Port Location Study (Tioga Group et al. 2001) was to analyse port related corporate activities and land uses. One major contention was to distinguish between those functions that definitely need to be located close to the port and those that might be dislocated to adjacent areas or even into the hinterland. Pulling out certain land uses that are considered not port dependent out of this "contested zone" would significantly improve the operating conditions of both the port and the related logistics service firms. Altogether with better connections offered by intermodal shuttle lines, the port would be able to further expand its container shipments, and the service firms would have more space for operation and less congestion that limits the efficiency of their businesses.

According to the Port Location Study that was published in 2001, and assuming a further annual growth of container shipments of about 5 per cent, the existing space reserve would have been running out shortly. Hence the Port Location Study had proposed a stepwise program of strategy development with different degrees of intensity and various requirements for implementation (Tioga Group et al. 2001). Speaking in general terms, the Port and the City of Oakland were advised to develop a joint strategy for implementation aiming at ensuring and further expanding the competitive position of the port, through an improved and more efficient use of space, a reduction of the negative impact of port and related operations exposed to community and neighbourhoods, and by the punctual dislocation of port related, yet not necessarily port (on-site) located services. There are several options for possible locations of satellite terminal and service infrastructures in the hinterland of the Port of Oakland suggested by the Port Location Study. Regarding the land side locations, these include commercial areas in the cities of Richmond and San Leandro on the East Bay with respect to waterway access, the City of Stockton in San Joaquin County had already been taken into account. The City of Stockton hosts and operates a bulk and dry port in the Central Valley, which also offers massive space for distribution centres. A rail link also exists that connects the Stockton-Tracy-area with the Bay Area. As named above, the co-operation that had been initiated with the City of Shafter (CA) and the MOL container line and others already aims at operating a satellite terminal for the Port of Oakland.

By the end of 2005, the Port of Oakland has also started an initiative for further co-operation with the Port of Sacramento. The short-term goal is to secure a terminal operator to provide a broad range of maritime services in Sacramento, which could offer services for customers of the Port of Oakland and thus enable the port to expand its market position. In the mid-term of this partnership, that is planned to run for about 10 years, it is intended to improve the conditions of both ports, e.g. by extending management and services, improve infrastructure conditions and accessibility. In the long-term phase of the co-operation, an exclusive Terminal Operations Franchise at the Port of Sacramento is considered. The Ports of Oakland and Sacramento plan to jointly develop a methodology to assess their performances.

However, even if extending its reach much beyond the immediate area of the Port of Oakland the port needs to increase the efficiency of land uses. The most critical point of the new locational policy suggested by the Port Location Study will be, first, to carefully assess the pros and cons of the existing land use structure and to prepare decisions where to continue use permits and where not. Second, the question will be how to convince those firms to move their location toward more distant areas that are considered being not necessarily linked to the port. Given a "zero-scenario", which assumes that no measures are being undertaken, an increasing number of firms will be pushed out of the core Bay Area (Tioga Group et al. 2001, 18). The main rationale for this expectation is, first, the general land rent differential; second, the disturbances for urban neighbourhoods caused by trucking operations; and third, traffic congestion as a consequence of the growth of consignment transshipped through the Bay Area. Insofar the region is urged to seek for alternative options for land use. Commercial areas dedicated to General Industrial Land Use in areas adjacent to the port are considered to be most appropriate for this aim. The Land Use Element of the General Plan of the City of Oakland makes respective suggestions for distribution and logistics related functions.

There are four different areas that are considered being potential alternative locations, according to the Port Location Study: i) The core "Port area", with the sites of the Port of Oakland, the Union Pacific Railyards, and a parcel of West Oakland; ii) The "Airport area", consisting of Oakland International Airport and the surrounding areas; iii) An area marked as "irregular" along San Leandro Street, about 10 kilometres southeast of the central port area; iv) A second area alike, along San Leandro Street, about 13 kilometres southeast of the port area, here designated as "San Leandro Street GIT South".

The potential for developing areas dedicated to port services and functions on the site of the port or the airport seems to be limited in theory, even if both areas provide respective space. In practice, the airport is already subject to expansion plans pursued by the Port Authority (that owns both entities, the Airport and the Port). However, there is are fact only small land parcels available for re-development. These land parcels do not really fit for the location of port services, since the lots are too small for the desired size and are also yielded by airport related users. Only those functions could be realistically located there which are suitable for the less space extensive air-freight services (e.g. freight forwarders that route consignments through the sea and air modes). The two other areas dedicated to General Industrial (and Transportation) land use, San Leandro Street North and South, are the only industrial areas that offer additional space for future demand. They comprise a gross-size of about 500 acres, and the existing building block of warehousing space comprises about 8.2 million square feet. These two locations for industrial and warehousing land use are the only commercial areas in Oakland that appear suited for a long term placement of port related functions. Some of the firms that are considered an appropriate target group with strong connections to the Port of Oakland are already located there. They are competitors of the port as well with regard to the question of where to acquire further land resources and their factual use. Most of this commercial land is already under use, however. In the case of off-site neighbourhoods, additional buffer space and measures to reduce related noise emissions are required. As soon as such spaces are becoming available, the port is advised to develop a mobilization strategy in order to ensure these areas for own use (see Table 5.8).

The General Industrial (and Transportation) areas in East Oakland, along San Leandro St, will presumably remain under this dedication for the coming years. They offer a significant potential for the location of port related or complementary services. This potential may only be mobilized if these areas are being used in an acceptable manner, judging from the community and neighbourhood perspective, and also if the related traffic and accessibility problem is being solved. In this respect, the designated rail-shuttle or a limitation of the truck through-traffic is

Table 5.8   Activation of port related land uses, in acres

	2000	2005	2010	2015	2020
Port owned land reserve	125	75	50	25	
Port services owned land reserve		75	75	75	75
Port expansion		15	15	15	15
City expansion		15	15	15	15
Total	125	180	155	130	105

Source: Tioga Group et al. 2001, 231

being considered. Insofar, the container port and the port related commercial areas in Oakland represent the typical problems of logistics spaces in an old-industrialized urban area. A dense built environment, an increasing locational competition and the squeeze-out of disturbing land uses are characterizing such areas. Regarding the planning and zoning policies of public sector, most Bay Area communities resist against what they see as trucking activity: "This is a major problem in locating any firm that ships or receives goods in any quantity. Every city wants *Microsoft* or *Pixar* to locate there, but there are not enough "clean hands"-firms to keep everyone employed and even those firms must eventually ship or receive something." (Freight Consultant, Interview)

This statement is very much in line with the municipal perspective:

> The City of Fremont does not track locations of warehousing and distribution facilities. With the exception of three distribution firms, it is not a major business sector within the City. Since we do not track this industry sector, we cannot tell you how many firms have left the City over time. The best we can say is as rents go up, this type of operation tends to become less viable. Fremont is very much part of Silicon Valley with hundreds of High-Tech firms. As rents have increased in the rest of the Valley, Fremont has become very attractive to High-Tech R&D uses, most recently bio-tech and telecommunications. This has caused Fremont rents to trend upward over time, so firms that have their sole function as warehousing or distribution have become less economically feasible in Fremont. (Economic development manager, The City of Fremont)

In summarizing the situation in the East Bay as an old industrialized corridor with a traditionally high attachment to logistics and distribution, this land use profile is increasingly becoming critical. On one hand, the proximity to the Port of Oakland and the high number of service firms that are located in the vicinity of the port offer a broad range of advantages for firms and customers in the East Bay. This is mainly derived from time and cost savings, due to the high accessibility of the majority of customers in this core area of the Bay. This is particularly the case in those subareas that are not under the pressure of high land rents (as e.g. the South Bay Areas is). There supply of warehousing and trucking space provided by the Cities of San Leandro and Hayward, in addition to the City of Oakland, is still high. On the other hand, empirical evidence is provided by employment data and real estate market data that the significance of the East Bay as a distribution space is decreasing over time. This is mainly caused by the shift of firms out of the core East Bay Area toward those regions that offer cheap land at lower cost, and also improved traffic conditions and accessibility. Such suburban distribution complexes have already emerged to a significant extent in rural counties adjacent to the East Bay. This is particularly true in the case of the Central Valley, which is presented in the following subsection in more detail.

b) Regional distribution complexes in suburbia: The cities of Stockton and Tracy, CA   Historically, suburbanization in the Bay Area had occurred in several waves of de-concentration depending on particular growth pressures and available space where businesses could be moved to (see above). As a consequence, those locations had increasingly been chosen which were located more distant from the core region, yet that were also well accessible from the urban centre via the freeway network. In the 1970s and 1980s, suburbanization was driven into the so-called Tri-Valley, a corridor along the eastward freeway I-580 that connects the Bay Area with the Central Valley, and the intersection with Interstate I-680 that leads to the North of the East Bay. Along these two corridors, service oriented business and commercial areas had become developed attracting firms that moved first into modern suburban business parks (Lang 2003, 73). According to Lang (ibid.), cities such as Pleasanton, San Ramon or Walnut Creek along the I-680 were thus representing the only de facto Edge cities of the Bay Area, except a particular location north of San Jose in Santa Clara County.

The spatial shift of service firms towards the locales of the second and third wave of post-war suburbanization reveals a change of urban functions and locations that did not only occur in the U.S., but also increasingly in Europe. As a consequence, a particular spatial division of labour has emerged regarding function and location, between core cities and the increasingly enriched periphery – also in those economic subsectors that were considered being mainly attached to core city areas. This was particularly the case with business services. Today the Tri-Valley that is located adjacent to the East Bay Area has been chosen by modern technology and service firms, e.g. Dublin and Pleasanton. Dublin, also accessible by the BART-Metro, big-box retail and leisure land uses have been established (e.g. a multiplex theatre). Meanwhile, locations like that are not affordable for transport, warehousing and distribution firms anymore. Consequently, a shift towards these places close to the core Bay Area cannot be taken into account for the logistics business.

These firms need to move far beyond, toward more remote locations. Since the freeway-network builds the time-space framework for such a spatial behaviour, the related pattern reveals the different locational opportunities. According to a detailed assessment by Oakley Strategic Economics Consultants (2001), four important corridors in different distance to the East Bay Area can be distinguished:

- Highway 4, stretching south of the Suisun Bay and connecting northern Contra Costa County with the East Bay,
- Interstate I-680, leading north-south across the East Bay, parallel to Interstate I-80,
- the core Tri-Valley at the intersection of Interstates I-680 and I-580, also
- the Tracy area in the Central Valley, located at the intersection of Interstates I-580 and I-5, the latter being the major Central Valley north-south arterial.

The two prototypical "edge city"-locations at I-680 and in the Tri-Valley are particularly remarkable due to their high share of corporate business services (42.3

per cent of the employees and 37.2 per cent, respectively). The corridor stretching along Highway 4 is characterized by a relatively even economic profile; it hosts the highest share of retail businesses among the four locales compared (24.4 per cent), and also a high score regarding construction (12.6 per cent). In contrast, the most distant location in the Tracy area is characterized as follows: it reveals the highest share of agricultural production (11 per cent), which was the traditional economic base of the Central Valley: it is even more occupied by manufacturing compared to the three other locations (14.2 per cent), and it ranks highest in terms of logistics, warehousing and distribution, four times higher than the other locations (12.4 per cent). However, Tracy is the least significant in terms of corporate business services (7.3 per cent of the employees). Tracy's economic profile appears to be more traditional, with high scores for agriculture and manufacturing, and also a disproportionately high volume of employment in transport, warehousing and logistics.

The real estate market confirms the economic profile expressed in the employment data. Due to the land rent gradients, new regional distribution complexes have been established in a distance of about 90 miles (144 km) from the East Bay, rather than in more proximate locations. The axis that stretches along the Interstate Freeway 5 between the Cities of Stockton, Tracy, or Manteca, distribution represents up to 80 per cent of the investments in the newly established commercial areas. This extremely high share seems to be a general property of places in San Joaquin County, and it underscores the importance of this particular segment of the Central Valley for the Bay Area. "Within San Joaquin County's 76 million square foot market of industrial buildings, approximately 79 per cent are warehouse and distribution, 19 per cent manufacturing, and 2 per cent research and development space." (San Joaquin Partnership 1999, 6) Among these firms, the share of corporations that had moved out of the Bay Area is supposed to be significantly high, according to real estate experts. Particularly large-scale distribution centres with a high demand for space may favour the cheap land rents in San Joaquin County. "Typically, the activities in warehouse space require large amounts of space and have relatively high space requirements per dollar value of goods/services involved." (Hausrath Economics 2003, 10) In the year 2003, the Average Asking Rents for warehousing were in a range between $0.74 per square feet in San Mateo County, $0.37 per square feet in the East Bay (I-80/I-880), and $0.30 per square feet in San Joaquin County (Hausrath Economics 2003, 9).

A significant portion of the commercial areas, where these new distribution centres have been established, can be found in the City of Stockton in San Joaquin County, namely in the "Airport Gateway Center", in the "Central Valley Industrial Park" and in the "Arc Road Business Park". The City of Stockton also hosts an inland port with a major land reserve and some old-industrialized lots as well. Also important in this respect is the City of Tracy, in the southwest of Stockton with about 30,000 inhabitants. One of the major commercial areas with a high attachment to logistics and distribution is the "Northeast Industrial Area", also the "Patterson Pass Business Park", situated on an unincorporated parcel of land west of the City. These two areas had been developed on merely open space in the 1990s. Whereas the former represents the regular extension of the urbanized area of the City of Tracy, the latter has been developed outside the built environment, yet with freeway access close by. With a size of about 800 acres and 600 acres respectively, they belong to the largest commercial areas of the entire region. The previously open space of a significant size seems to be prototypically appropriate for the purpose of hosting distribution with its flat, huge buildings. Altogether with the warehouse buildings that had been established in the Tracy area before, a significant contribution to regional economic development has thus been made (see Table 5.9).

Table 5.9   Distribution firms/locations in Tracy/CA

Firm	Type	Employees
U.S. Defense Logistics Agency	Defense logistics	2,067
Summit Logistics	Grocery distribution (Safeway)	1,500
Costco Wholesale	Grocery and Non-Food-Distribution	470
Yellow Freight	Freight forwarding	300
Orchard Supply Hardware	Household equipment	300
DSC Logistics	Freight forwarding	100
U.S. Cold Storage	Frozen distribution	
Office Depot	Office supply	n.a.
United Grocers	Grocery distribution	n.a.

Source: own investigation, 2001

Microperspectives

The following section summarizes the findings of a qualitative survey that had been conducted in 2001, mostly consisting of expert interviews with corporate representatives in wholesale and retail distribution and also in logistics (total 26 firms). The firms were located in the Bay Area and in the Central Valley). Also, 21 experts from the areas of urban planning, commercial development and the real estate business had been surveyed (see Figure 5.2).

a) Corporate decision making   The locational decisions made by firms about their goods distribution facilities are based on three main issues: i) land cost and land competition; ii) strategic transport access (long-distance/regional) and accessibility to customers; iii) and, particularly regarding the more distant

SPATIAL ORIENTATION

LOCATION	Oakland	East Bay Area	Central V.	Large-scale
Oakland	x	x		x x
East Bay Area	x x	x x x x x x		
Central Valley			x x x	x x x x x x x x x x
Other				x

Figure 5.2   Cases of firms studied in Northern California

Source: Own

locations, also affordability of the workforce, both in terms of low wages from the firm's perspective and the low cost of living in an area for the employees. These factors, that cover most of the answers of the firms that were surveyed, are to be balanced carefully, which seems to be independent of where the company locates: in the core East Bay Area or in the Central Valley. Given the historical background and infrastructure development in the East Bay, the main location factors explaining the density of goods distribution facilities in the region are as follows: the presence of a major port, an airport and the interchange of two major highways, namely the I-880 along the East Bay and the I-580 connecting the Bay with the Central Valley. Good access to the Peninsula, and to South San Francisco via the San Mateo Bridge also plays a role, crossing the I-880 in Hayward. The disadvantages of this location relate to the population and land use density of the region, the related transport constraints, and the fact that the freight business is regarded as an undesirable land use.

As the main rationale for the establishment of distribution centres on the large, newly developed lots of the Central Valley can be identified by the large and cheap parcel of land, the freeway-access close by and the strategically favoured location close to the Interstate I-5, which allows for to provide goods distribution for areas much larger than just the northern Central Valley or the Bay Area. The more peripheral location allows for mobilizing locational advantages primarily against the background of inter-regional distribution areas, compared to places in the East Bay Area. Distribution is then being organized in a spatial stretch including Southern California or even beyond. In some cases, rail freight-access seems to be important as a location factor as well. According to a representative of the City of Tracy, the locational setting that drives firms over here appears quite clear:

I am not sure if this is too simplified ...: circulation; availability of rail lines, freeway access, proximity to ports and to airports have been an influence. Real estate prices in the Bay Area have also been a major factor in the increase of Tracy's appeal. High real estate values due to limited development space in the core Bay Area have resulted in a focus on outlying areas, included the City of Tracy. What separates Tracy from other Central Valley cities is basically its proximity to the Bay Area. (City of Tracy, City Planner)

Particularly the Interstate I-5 is increasingly developing as the backbone of California, since distribution from places along this corridor may cover both Northern and Southern California. Many of the logistics firms tend to operate one DC in each of the two major parts of the State, thus providing the entire California population of about 34 million consumers. The related ratio between the number of facilities and the size of the catchment area offers a tremendous mobilization of economies of scale just by the concentration of DC-functions and the related rationalization of operations. The Patterson Pass south of Tracy, with its size of about 600 acres, represents a typical example of these large-scale distribution complexes (see Map 5.3). It includes the DCs for the northern California distribution of a major grocery retailer and a club-like wholesale trader, who are serving their Central Valley and Bay Area outlets out of these DCs. The grocery-retailer's DC

Figure 5.3   Distribution centres in the Patterson Pass Business Park, Tracy/CA

Source: Own

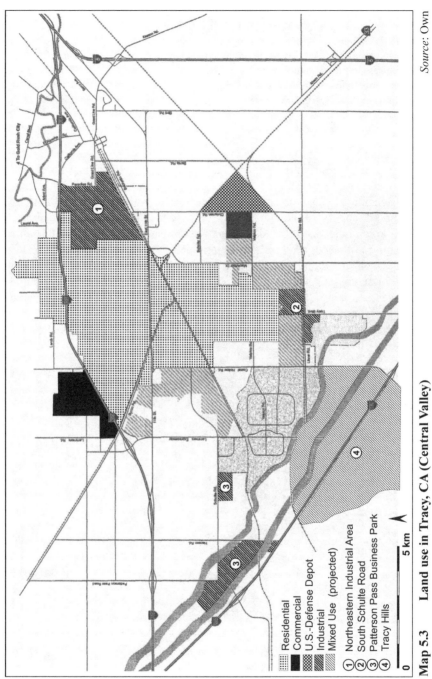

Map 5.3　Land use in Tracy, CA (Central Valley)

Residential
Commercial
U.S.-Defense Depot
Industrial
Mixed Use (projected)

① Northeastern Industrial Area
② South Schulte Road
③ Patterson Pass Business Park
④ Tracy Hills

0   5 km

Source: Own

comprises a warehousing space of about 2 million square feet and represents one of the biggest single DC-facilities in California (see also Table 5.7).

The single most important issue regarding corporate locational behaviour seems to be geography – the macro-location in the context of the entire region and against the background of the accessibility of the customers. Geographical presence is being mediated by the factor transport accessibility, which may influence the value of the macro location either in a positive or critical way. The related assessment of the interplay of mobility and immobility tends to be included in almost any firms' decision when seeking for location. This underscores the still vital role that is being played by the factor "space". It appears to be predominant for this group of firms or sector of the economy, even in the age of the Internet and accelerated technological and structural change. Even more: locational choice and the optimization of mobilities seem to be key issues for firms that are parts of broader networks, adjusted to the management of space and time. The supply of properly sized lots ranks next in corporate assessment, combined with land rents as the thus most important criteria that highlights places in the Central Valley. The land reserves of the Central Valley and the respective, "robust" environment of the facility (e.g. no restrictions for 24/7-operations) rank second from the firms perspective, as an incentive to leave the Bay Area in favour of a Central Valley location.

However, among the firms surveyed in the Central Valley only a part had been moving from the Bay Area directly (five out of 13 firms interviewed). The remaining firms had been newly established from corporations that are based outside the region: in order to open up a new business, or to expand and move to the Central Valley, by keeping the old locations running. This is e.g. true for couriers and messengers, express- and parcel-services that are still present at all micro-locations of the Bay Area. They have been using the Central Valley as a place for business expansion and extension of their distribution area, rather than to replace core city locations with more remote places. It is unclear whether such a locational split can be profitable. This may depend upon the density of distribution points: the higher the density of destinations is, the higher might the productivity of distribution become; however, the more complicated may the logistics operation become as well. Such companies then seek for compromizing between the imperative of proximity to customers and the complex operational conditions for distribution in core city areas. Corporate practice in the Central Valley seeks for optimizing both dimensions. The resulting consequence is to place the DC as close to the customers as possible and as remote to the areas of high land rents as necessary. In this respect, the locational strategies of firms in Northern California and in Berlin-Brandenburg appear to be comparable.

A third major factor that had not been mentioned in most of the expert- and corporate interviews in Germany belongs to the labour market. Different from the Berlin-Brandenburg region, labour resources score extremely high in the entire Northern California area, particularly in the Bay Area. The availability of workforce particularly at the lower and medium levels of warehousing and trucking appears to be an important constraint for developing the distribution economy further. Northern California might not be affected by labour shortage to the extent that had occurred in Los Angeles and Long Beach, where the San Pedro Ports experienced such related excessive delays in freight handling in 2004. However, transport and warehousing firms in the entire Bay Area report that – besides the land factor – labour represents the most important constraint for their businesses. In turn, all firms surveyed in the Central Valley did mention this issue as a particular competitive advantage of their region.

Similar to the German case study, core city- and periphery-oriented firms can be distinguished as ideal types, since they are characterized by distinct differences in their locational preference. Each is specifically organizing the respective interplay of fixed costs (land site), variable costs (transport, labour) and political regulations that may explain a majority of the site-selection decisions. A third group appears different: the users of intermodal services. They are of course location-wise depending on the supply of infrastructure, particularly the interface between the different transport-modes that is provided either in a port, freight railyard or airport. Whereas the Berlin-Brandenburg sample included just one provider of rail-freight services, these services are much more frequent in Northern California – both from the supply and the demand perspective. This has to do with the container shipments handled in the Port of Oakland and also with the freight segment at the San Francisco, Oakland and San Jose Airports. Their services certainly attract related land uses and, due to the high degree of time sensitivity of shipments by air, the latter firms need to stay close to the airport. However, even in this market segment, tendencies of spatial expansion and shift (e.g. of terminal infrastructure) can be observed. They are caused for mere space reasons and also reflect rising traffic problems in the congested areas of the urban core.

b) Cities and regions   Cities and regions as the second major players in this respect are often critical of freight transport. Municipal strategies will further restrict the goods distribution businesses, since those firms are low tax-payers, have relatively few employers, and – certainly of major concern – are important generators of truck transport, which often adds to local congestion. However, this attitude appears to be a matter of "affordability": It can be practised as long as there are better alternatives available. Once the traditional industries are declining, due to structural change and globalized competition, and the new high-tech sectors prefer other places, for whatever reason, there may be no alternative rather than to seek for transport and warehousing firms, judging from a municipality's perspective. Whereas all cities surveyed in the Bay Area sought service activities, high technology and light manufacturing firms, the planning attitudes in the Central Valley appear different, owing to the large amount of DCs being established there. However, even the current winners of the outward drift of distribution into the Central Valley confirm, at least officially, that they would not deliberately seek additional goods distribution firms. Although, in the case of certain interest, they will certainly agree and attempt to realize such investments, since most of the municipalities may need to strengthen their tax base.

One industry that was *not* selected as a "target" is targeting San Joaquin County: distribution and warehousing. Seeking immediate highway access and other critical transportation infrastructure, central location to the markets they serve, low land costs and/or relatively low rents, large footprint "big box" users are focusing on San Joaquin County. (San Joaquin Partnership 1999, 6)

The City of Tracy provided another good example. Although city officials claim to be interested in higher quality investments, the city administration still promotes the area as an attractive place for goods distribution firms. One of their official marketing tools is a video that cleverly announces: "All Roads Lead to Tracy".

Indeed, in general, the surveys revealed a mixed picture. On one hand, most cities do not actively discourage the addition of new goods distribution firms, may be except of centres such as Fremont that are close enough to the Silicon Valley and will in fact be targeted by the desired high technology firms. It is also evident that even high technology sectors depend upon an efficiently working freight system. On the other hand, neither of the municipalities has developed strategies in order to cope with the growing freight traffic that affects the region. If any attention is paid to the freight transport sector, it is carried out by widening the infrastructure and sometimes by promoting rail and water transport to make the system run more efficiently. Concerns about sustainability with respect to truck traffic, the effect of negative externalities and attempts to provide better acceptability for freight transport do not seem to be the subject of concern or policy development in the region (on this issue, see also the next chapter on policy and planning strategy).

In addition to the development of commercial and industrial land use, Central Valley locations are also characterized by the designation of housing areas. In this respect, they aim at both offering workplaces and housing opportunities for target groups that are supposed to move out of the Bay Area. Instead of attracting reverse commuting, and in addition to jobs offered at the entry-level and higher (e.g. warehousing, trucking), they are also massively investing into the housing real estate market. In light of a major problem of the Bay Area – the scarcity of housing space for low-income households, due to the high land rents and the high cost of living – the small towns try to present themselves as an attractive alternative, either to the high expenses of living in the Bay Area or to the widely spread commuters' lifestyle. In the City of Tracy, an average single-family house is available for about US$200,000–220,000, which makes them affordable for a major part of low-income and middle-class communities, whose salaries paid by distribution businesses vary between US$8.50 per hour (warehousing) and US$19 per hour (trucking, sales). To earn a living based on this income appears to be almost impossible in the Bay Area.

Chapter 6
Logistics and Freight Distribution from a Policy and Planning Perspective[1]

A Freight and Logistics Policy Framework at Metropolitan and National Levels

The enormous problems associated with the growth of freight shipments and the related transport needs have triggered both capacity and acceptability constraints of the current distribution system, of which the former is generally accepted as a serious challenge to policy and planning. In contrast, sustainability of freight transportation is (still) subject to minor consideration, because economic interests are often ranked much higher than social or environmental goals used to be (Black 1996, 2001). However, judging from the perspective of policy and planning, freight transport and logistics is an increasingly important issue, and it also represents a target difficult to manage. This is due to the cost-sensitive character of freight transport as subject to corporate management and decision making, which is different from passenger transport, where decisions are mainly made by individuals, following more than just cost-based rationalities. Freight is both an outcome and a component of highly abstract network architectures that are not necessarily open for external management, for example, for governance in the public interest or in response to local issues. Freight transport remains in private interests that seek to maximize system-wide utility. Finally, the potential degree of any planning intervention depends upon the regulatory framework which has been changing significantly over the last two or three decades, thus driving freight growth through shrinking barriers for trade and transport, falling freight rates and a highly competitive environment in the logistics service industries.

Once taking a closer look at the regulatory framework and the physical operationality of the freight distribution system, the current situation appears quite contradictory with de-regulation and market liberalization on one hand, in order to allow for accelerating freight flows, and increasing constraints due to infrastructure bottlenecks, urban density and scarce land on the other hand. As a consequence, there is a remarkable contrast between the fluidity of flows and the inertia of the physical infrastructure, even if one acknowledges the rising significance of information flow and managerial competence. Because transport systems, particularly infrastructure and land supply, cannot accommodate the

---

1 This section makes particular reference to issues developed in close collaboration with Jean-Paul Rodrigue (cf. Hesse and Rodrigue 2004, Rodrigue and Hesse 2007).

growing amount of freight traffic, the question is how the associated problems might be solved in future, with much higher transport volumes in addition to the performance of the current systems.

To answer this question, it makes sense to look back and raise the issue of how municipalities and transport planning authorities have been coping with these problems in the past (see Banister 2002). In general, transport planning has for a long time been focusing on managing passenger flows and did not extensively develop plans and strategies for freight distribution. In many cases, logistics and freight distribution were considered an undesirable land use at the local level, at least in economically prospering regions (in others, logistics firms have been welcomed for the sake of certain economic benefits, such as jobs, local tax revenues, etc.). Planning activities with respect to truck transport and rail freight have been undertaken only recently, compared to passenger transport and the respective tradition of modelling, traffic counting, etc. Regarding the way freight distribution and logistics have been covered by policy and planning in practice, different stages can be distinguished: During the 1960s, freight has not been particularly addressed by transport planners yet, except the matter of fact that port development in general represents a primarily freight related issue. Planning practice in the 1970s/1980s did indeed pay more attention to freight, yet mainly followed the traditional guidance of "predict and provide", focusing on measures that were devoted to widening and expanding the infrastructure network. Not earlier than in the 1990s, the issue of inter-modality evolved as a generally accepted paradigm for policy and planning. Whereas the de-regulation of transport markets has substantially lowered the degree of government intervention, to some extent air quality policies have been introduced as new regulation tools, for example, addressing emission standards.

In North America, a substantial increase in freight related activity at both metropolitan and national levels can be noted, starting at the end of the 1990s and early 2000s (Rodrigue and Hesse 2007). According to the accelerated growth of freight transport and the rising degree of conflict, urban economists, transport planners and the trade sector share a rising interest in freight issues (Eno-Foundation and The Intermodal Association of North America 1999). This was introduced in order to make freight and logistics more efficient and more acceptable, by integrating freight into planning schemes and frameworks and also by offering training and education programs.

Regarding both capacity and sustainability constraints of the current freight system, there is a need for developing a balanced framework of policy and planning measures that consist of more than just adding to infrastructure. It comprises generic policy approaches (with respect to energy, climate change, infrastructure policy and modal share), inter-modality as a key tool, and also balancing the freight sector with community demands, for example, with respect to traffic generation or neighbourhood issues of inner-city distribution centres (cf. TRB 2003). Regional examples such as the Seattle/Tacoma "FAST Corridor", the Alameda Corridor or other initiatives in metropolitan regions underline attempts to try to divert freight in a firmly established national trucking market. Although on paper these initiatives appear quite reasonable and promising, the existing distribution system takes time to adjust. Therefore achieving the modal shift they were designed for may take much longer than expected, whereas in the meantime road freight transport is growing further. Case studies may even provide evidence to suggest that attempts at freight planning are not that useful, unless being developed by the private sector or at least in close cooperation with it. For example, the Port Inland Freight Distribution Network of the Port Authority of New York and New Jersey has also shown a rather slow start with much less traffic than expected in spite of subsidies and incentives. Thus modal shift strategies, either planned or left to market forces, are facing substantial inertia reflecting accumulated investments, routes and management practices.

In Europe and Germany, planners and policy makers have started to develop a policy agenda focussing on freight transport by the end of the 1980s and the beginning of the 1990s. This was not coincidentally linked to the introduction of the Single European Market which was expected to be followed by a significant increase in overall transport volumes, particularly with respect to international freight flows. Since then, European and national policy programs were focussing on improving alternative transport modes, such as rail and barge transport, also combined traffic and intermodality respectively, both with emphasis on routes within networks (e.g. the Trans-European Transport Networks TEN), and on nodes and terminals. However, as ambiguous as transport policy tends to be, the outcome of such initiatives appears to be limited, for several reasons. First, there is a strong economic rationale behind European transport policy that considers accessibility an important means for generating wealth and, particularly, cohesion. Thus European regional and structural funds aim at expanding transport networks, which are focusing on the road transport mode to a considerable extent. Second, there is a clear contradiction between the overall policy goal of de-regulation and liberalization (of goods and services markets, particularly of transport markets) and the regulatory requirements that may be considered essential for achieving policy goals such as a significant modal shift. Third, as outlined below, logistics and freight distribution is not only a contested yet also complex issue that eludes from intervention to some extent. This had already become evident once the Federal Government in Germany established its policy to support the introduction of integrated freight centres in the early 1990s, yet failed in influencing the modal share in freight transport to a significant degree. The wisdom that logistics works as a tool to integrate the highly fragmented, widely dispersed landscapes of manufacturing and distribution is challenged by serious conflicts both within the system of logistics and freight distribution and also with its environment. In order to properly reflect these properties, the concept of friction has been introduced as a major framework for understanding and better adjusting policy and planning to the complex subject of logistics and freight distribution.

The Concept of Friction as Means to Understand Supply Chain Conflicts, Transport Constraints and Planning Strategies

The concept of impedance, or the friction of space, is central to many geographical considerations of economic and social processes. Conventionally, this concept was subjugated to issues concerning distance and how to quantify it. Substantial economic research has focused on assessing impedance, the impacts of distance, time and elasticity on freight flows (Button 1993). As discussed so far, significant changes have incurred in freight transport nodes, flows and networks, which impacted on the concept of impedance. Logistics and freight distribution, as a transport paradigm, require a review of this multidimensional concept to include four core elements, namely the traditional transport costs, but also the organization of the supply chain and the transactional and physical environments in which freight distribution evolves. These four elements, which are difficult to consider independently, jointly define the concept of logistical friction and its possible improvements.

The first element in the concept of friction relates to transport and logistics costs. Traditionally transports costs were considered as a distance decay function. The most significant considerations of logistics on transport costs are related to the functions of composition, transhipment and decomposition, which have been transformed by logistics. More specifically, composition and decomposition costs, which comprise activities such as packaging, warehousing and assembly of goods into batches, can account to 10 per cent of production costs. A higher level of inventory management (e.g. lean management) can lead to significant reduction in the logistical friction as well as terminal improvements decreasing transshipment times and costs. Time is becoming as important as distance in the assessment of transport costs and impedance. As transport costs went down through space/time convergence: the value of time went up proportionally.

However, within the components of logistics costs, the transport segment has experienced absolute as well as relative growth. Inventories are increasingly in circulation, and inventory costs were reduced proportionally. The issue of mobile inventories, as opposed to the traditional concept of fixed inventories, has blurred the assessment of logistics costs. Trade-offs between fixed costs (inventories, warehouses etc.) and variable costs (transport) play a major role in corporate strategies, since the advancement of new technologies allows for the mobilization of inventories and, subsequently, the elimination of facilities – whereas the deregulation of transport markets attracted firms to expand their shipping and transport related activities by significantly lowering the freight rates. Thus companies were able to reduce a considerable amount of total distribution costs.

Complexity of the supply chain comprises the second issue of friction that deserves consideration. An integrated freight transport system requires a high level of coordination. The more complex the supply chain, the higher is the friction because it involves both organizational and geographical complexity. Under such circumstances, the logistical friction takes the form of an exponential growth function of the complexity of the supply chain. A core dimension of this geographical complexity is linked with the level of spatial fragmentation of production and consumption. Globalization has thus been concomitant with a rise in the complexity of the supply chain and logistical integration permitted to support it. Many industrial location concepts indirectly address this perspective by investigating how firms grow in space and how production is organized to take advantage of comparative advantages (Dicken 1998). The extended range of suppliers and the globalization of markets have put increasing pressures on the supply chain: a problem partially solved by using high-throughput distribution centres.

The geographical scale of the supply chain is linked with a level of logistical friction as nationally-oriented supply chains tend to be less complex than multinational supply chains, mainly because they are less spatially fragmented. From an operational perspective, it considers a balance between the benefits derived from the increased fragmentation of the supply chain with the organizational costs that come along. At some point, it becomes excessively difficult to maintain the coherence of the supply chain. The marginal costs of this function have substantially been reduced by information technologies and corporate strategies such as mergers and joint ventures implying that increasingly complex supply chains can be supported with the resulting improvements in productivity, efficiency and reliability. Consequently, it is possible to maintain or improve key time-dependent logistical requirements over an extended geography of distribution, namely the availability of parts and products, their order cycle time, and the frequency, on-time and reliability of deliveries. The consolidation of logistical activities in high-throughput distribution centres or the reliance on third-party logistics providers (3PL) (which are using their own terminals or DCs) are strategies to reduce the friction of the supply chain.

In terms of both competition and co-operation, the transactional environment within which logistics is performed is becoming increasingly important, since the fragmented organization of value chains inevitably raises the issue of transaction (Visser and Lambooy 2004). According to the challenge of logistics integration, a rising number of firms and locations are bound together in material flows, chains and networks. In order to operate their businesses efficiently and competitively, these firms establish complex relationships that are performed by contracting (vertically or horizontally), by competition (horizontally), or, in rare cases, by co-operation. Since different players pursue different interests and have distinct authority to realize their vital interests, their transactional environment is characterized by structural tensions. They unfold distinct power issues along the chain, which is regarded as a major source of logistical friction. This issue has extensively been discussed – with regard to producer-supplier relationships – in the automotive industry, where the costs of time, of uncertainty and risk (notably expressed in transaction costs) were passed on to suppliers and subcontractors. Regarding the widely practised outsourcing in distribution and logistics, such behavior is becoming more and more common in this business as well.

Supply chain power is particularly performed by firms who are acting as purchase and order agents, e.g. large retail chains who are buying transport services from 3PL logistics firms, freight forwarders who are trading and brokering orders, large ocean shipping companies who are responsible for moving a considerable amount of cargo worldwide, or large conglomerates having multiple production and distribution units. These units are able to command the conditions of delivery that have to be fulfilled by service providers. In order to cope with this pressure, transport and distribution firms are impelled to provide high service quality at low cost, in an increasingly competitive environment. The uneven distribution of power is primarily due to specific supplier-customer relationships, it depends on the firms' position within the chain, related to market demand (Taylor and Hallsworth 2000), to its organizational or technological know how, or to factors such as the mere size of the firm. In comparison to traditional approaches mainly based on transport costs, the concept of logistical friction mirrors a more comprehensive understanding of the constraints and capabilities of firms.

The transactional environment also includes regulation issues, since public policy appears as another major factor of influence. Even in the age of transport market deregulation, government issues and the public sector remain influential on the distribution framework and thus on the freight traffic performance. This is due to legal issues, particularly the enforcement of load and vehicle inspections, to the continuous check of drivers' working hours or to the definition of vehicle noise and air quality emission standards in order to improve the environmental performance of truck fleets. At the local and regional levels, zoning policies and building permits regulate the locational constraints and opportunities of the firms. To some extent, firms depend upon the establishment of a co-operative transactional and regulation environment – even if capital nowadays seems to be more powerful than the state or the local communities appear.

Finally, the physical environment of logistics and freight distribution has to be taken into account as a major framework condition for operating logistics and solving associated conflicts. The physical environment comprises the "material space" where any social and economic activity is embedded in and also the hard transport infrastructure that is necessary for the efficient operation of the system, like roads, railways, warehouses, terminals or ports. Such physical environment appears as a major external determinant of the movement of vessels and vehicles. It thus can become decisive for the success or the shortcoming of the distribution system. Normally it is regarded as a component of transport costs, since infrastructure bottlenecks or road congestion do harm the firms" productivity in terms of delays and malfunctions. This follows the more critical consideration of space as a barrier for the seamless physics of flow.

In the concept of logistical friction, the physical environment plays a more sophisticated role since it represents the entire pressure that is exerted by space on the supply chain, positively and critically. This happens particularly in those areas that are characterized by scarcity of access: e.g. congested places such as port areas and port hinterlands, or core urban areas that are problematic for delivery.

They are not only bottlenecks for exact channel distribution, but also traditional locales where logistics is committed to adjust to its built environment. This is the simple reason why urban delivery vehicles are lighter and smaller than the long-distance trucks. Port areas embody the contradiction between scale economies and the limitations of infrastructure and facilities in a very typical way: in cases where simple expansion of a port system is out of question, due to space, money or policy constraints, the agents of distribution have to arrange themselves with their environment. Such ability to balance different interests, originally caused by constraints for the usual path of development, can also be considered positively, as a source of creativity and innovation.

The concept of friction delivers a better understanding of the often constrained and contested environment within which logistic and freight distribution systems are being operated. Against this complex background of corporate strategies, competitive and co-operative interaction on one hand, and also physical and environmental conditions on the other hand, policy and planning activities with regard to logistics and freight distribution have to be carefully assessed and implemented.

The Berlin-Brandenburg Planning Regime and Strategies

The Berlin-Brandenburg region is part of a complex, multi-layer system of political decision making and planning practice, where different levels and authorities come into play: first, Federal policy, setting the legal framework and also directing a certain portion of the required money flows, due to the funding authority of the Federal government; second, State policy is highly involved both in Berlin and in Brandenburg, with Berlin representing the special case of the "Stadtstaat", a mixture of State and municipal authority; regional policy and planning, particularly responsible for the zone of mutual interdependence, already introduced as the surrounding area or "suburbs"; and finally the local level, both in Berlin (mainly organized district-wise) and in the Brandenburg communities as well. The issue of this study is primarily subject to local and regional policy and planning. However, given this somehow unique regional framework, the State-level is particularly involved, since the strategic plans for sustainable urban development and related sectoral plans (such as for mobility) were released by the Senate of Berlin. Also, the concept for establishing integrated freight centres in the Berlin suburbs were an outcome of joint activities pursued by the two States of Berlin and Brandenburg from its very beginning in the 1990s.

In 2003, a comprehensive plan on urban development and transport ("Berlin Mobile 2010") had been released by the Senate of Berlin, the State and City administration which is the main political body of the City of Berlin. The concept was based on the premise that the efficiency of the transport system and the sustainability of urban development do not necessarily exclude each other, but do have a common ground where modern planning strategies can tie up to. As a part of this plan, also an integrated concept for managing commercial traffic had been launched in 2006 (Berlin Senate 2006). The aim of the concept is to ensure the functionality of all activities that are associated with goods movement and service related mobility. Also, the general aims of "Berlin Mobile 2010" should be supported by this partial strategy as well. To some extent, the freight distribution sector is conceptualized as a system with complex dynamics that need to be better understood than before. So, in addition to conventional measures such as infrastructure supply and traffic management etc., corporate strategies had been taken into account in more detail, both as a potential problem (e.g. just-in-time production modes) and as a possible solution (e.g. collaborative distribution modes). Thus the concept included an analysis of system dynamics, specific spatial contexts and regulation issues. In this respect, freight and logistics related planning is more considered a joint learning process of business and policy, rather than the mere implementation of planning schemes and measures.

The Berlin-Brandenburg Freight-related Policies – A Critical Overview

Initially, the Berlin-Brandenburg freight transport concept consisted of three items: the operation of the integrated freight centres that were already introduced earlier in this volume; the improvements of urban delivery conditions, mainly through local private-public co-operation; and the development of construction logistics for large building projects in the Berlin city centre, primarily operational during the 1990s with its extremely intensive construction activity. After the basic elements had been pursued in practice for about a decade, an initial assessment had been undertaken, regarding both the degree of implementation of measure and the potential impact these measures have had in terms of effects on either transport, land use or the environment, just to mention the most important dimensions of the concept (see Hesse 1998b, 2004).

According to this assessment, it can be confirmed that freight traffic and logistics in Berlin-Brandenburg generally benefit from an ambitious planning approach and advanced practical experience. Although the concept had been enforced without a broad database or a comprehensive freight transport model, the goals appeared to be solid for the improvement of organisation and acceptability of freight transport, at least in theory. The assigned instruments were state-of-the-art and even be convincing when judged from the supra-regional level (cf. European Commission 1998; OECD 2003). However, once summarizing the implementation record of the concept, there is an obvious discrepancy between the ambitious freight concept and a certain degree of implementation on one hand, and relatively modest effects in practice on the other hand.

This statement applies to the three components of the integrated freight transport concept in a different way. Among the measures implemented, only the construction logistics appears to be a success story, thanks to obvious benefits for public and private players involved. The supporting framework condition for this was the nature of the construction sites as both constrained environments (in terms of traffic accessibility) and attractive locations. This made corporate self-regulation and public policy successful, in order to introduce innovative practices and to mobilize certain benefits. Yet, the life-cycle of construction logistics appears to be short termed, as long as, first, it does not represent far more than a unique pilot project, and second, its main principle is not included in the regular planning scheme. In this regard it is meaningful that the well-known concept of construction logistics is currently no longer being applied at least not in Berlin. So the theoretical and practical benefits of this concept might not easily be generalized.

Contrary to construction logistics, the implementation of IFC and local delivery improvements in the Berlin-Brandenburg region still lacks significance. The reasons for these shortcomings are multifaceted and hence deserve a deeper reflection. They belong both to generic characteristics of the logistics system and to regional specifics, the latter explaining the degree of Berlin-Brandenburg related problems or failures. First, as mentioned above, there is the complexity of the entire logistics framework consisting of a variety of firms, strategies, products, services, and respective requirements. Instead of one or two major players responsible for a large amount of freight, a fragmented scenery of shippers, carriers, brokers, receivers is predominant, playing distinct roles in the supply-chain and pursuing individual interests. This multitude can hardly be harmonized or comprehensively managed. Even if certain modes of co-operative logistics organisation were successfully established: is it really worth eliminating "eight out of 80,000" lorry trips a day (as was once the case in the City of Cologne)?

Secondly, goods movement derives from processes of inter-regional exchange between enterprises and households. Logistics chains are often managed and controlled far away from the (urban) space that is affected by freight traffic. Cities and regions are sources and destinations of freight flows that are increasingly operated from remote locations with many overlays and subsystems. Based on history, this is particularly true for Berlin. Thus the limited power of policy and planning to manage freight transport reflects the ambiguous character of logistics as a driver and a mirror of the increasingly integrated spatial economy. A third major external factor is the rising competition in transport industries. Deregulation and increased international sourcing made the transport rates decrease (which is likely to attract further demand for transport and to diminish incentives for reducing traffic) and the shippers or freight forwarders reluctant to practise inter-firm co-operation. Site-selection and logistics operation decisions strictly follow cost-minimizing and efficiency concerns and often deviate from what planners use to desire (see Table 6.1).

In the case of the Berlin-Brandenburg region, as well as in many other European regions, the scope of policy and planning depends upon a particular climate of public and stakeholder awareness. In this respect it is notable that transport policy is increasingly perceived as – necessarily – business oriented. Measures that may restrict freight traffic, in order to achieve public goals, are highly unpopular raising fears of corporate exodus and job losses. In order to overcome this problem, the Berlin-Brandenburg model was based on a particular public-private partnership.

Table 6.1   Assessment of the integrated freight transport and logistics concept in Berlin-Brandenburg

Component	Advantages	Disadvantages/ Barriers	Effects in the region	General recommendations
IFC	Theoretical potential: - Intermodal access - Host disturbing, freight intensive land uses - Cooperation for load sharing and bundling	Time and money consuming, high public investment needed Individuality of corporate behaviour Competition in transport industry	IFC-participation of firms by site purchase, any other commitment IFC-share of inbound freight about 20% Site approach successful; no intermodality used; no cooperation	Develop long term goals and planning strategy Deliver incentives to firms (good product, access, synergies) Concentrate subsidies at preferred locations, limit concurring places
Delivery improvements (City Logistics abandoned)	Improving road and traffic conditions for delivery vehicles	Participation solely voluntarily, no accompanying planning activity	Only punctual, due to selective, voluntary participation No general traffic or environmental impact	Only reasonable in the context of comprehensive urban logistics, otherwise symbolic Urban logistics require active planning and communication
Construction logistics	Acceleration of building projects Shift from truck to rail and waterway Management/reduction of delivery trips by truck	High cost Expenses f. coordinating investors, builders, construction firms Infrastructure and interface development	At Potsdamer Platz very high (high share of rail and waterway, truck access restricted) Other cases with no impact (due to less or no regulation)	Manage large building sites at city centres w/ active construction logistics Extend construction logistics to the planning and organisation of regular delivery

Source: Own

Though regarding the "public" stake in this constellation, a more active role of its holders would be essential. Since this is normally associated with a rising degree of conflict, it turns out to be unpopular among city administrations who want to be consensus-seeking and business-friendly, especially since the modern economy appears to be increasingly flow-oriented. In this respect, the current role of public players in Berlin seems to be too reserved, too passive. A second regional factor is given by the fact that the Berlin-Brandenburg policy approach is mainly supply side-oriented (focusing on infrastructure provision) and thus increasingly neutral for corporate decision-making. Regional firms do not use the IFC intermodal facilities yet, at least not to the extent that was initially expected; this is due to higher cost and the lack of service provision. In this regard, the concept could be better accomplished by knowing more about the particular regional demand and by offering appropriate services.

Under these circumstances, one question is how to assess the collaborative approach practised in the Berlin-Brandenburg freight concept properly. Collaborative planning recently tends to be as popular as complex (Healey 1997). In fact, managing logistics and freight transport at the regional level – this is the lesson from the Berlin example – requires a development of a "hybrid" type of policy: A purely planning related discussion of this topic appears inappropriate, since it neglects major system dependent characteristics. Conversely, a mostly business dominated freight transport concept will probably mean that public policy concerns are not being considered. The optimum may be situated in between: in developing a comprehensive, both realistic and ambitious policy strategy for the freight sector that also receives strong support by State and Federal politics. Keeping in mind conventional thinking in transport planning, how can regional communities step farther beyond and be really innovative? Interestingly, it is the Berlin-Brandenburg arena that could be prepared for that kind of policy.

The freight and logistics concept has recently been updated in the context of the development of the "Integrated Concept for Commercial Traffic in Berlin" that was completed and published in 2006 (Senate Administration for Urban Development 2006b). This concept was based on a detailed analysis of current trends in economic and transport development. It also included an estimation of future framework conditions for related policies. Main results revealed that, different from other regions, future freight transport volumes are supposed to decrease in the Berlin area, whereas services related merchandise transport will further increase. In general, the demand for flexibility directed to freight transport carriers will favour the road traffic mode. However, there are new potentials expected for freight rail and barge transport modes through logistics innovations and new technologies, also due to increasing container transhipments. The concept defines measures to be implemented according to priority in several fields of action. First, those innovations have to be pursued or even expanded that had been realized in Berlin earlier: this applies to priority lanes dedicated to bus and lorry traffic, the communicative "Commercial Transport Platform", the establishment of particular parking zones for delivery vehicles, also the continuation of the co-operative construction logistics. Such ideas had already been proven in Berlin and are also subject to planning and implementation elsewhere in Germany or Europe.

Second, the concept includes issues that may ensure the sustainability of commercial transport in Berlin with respect to changing framework conditions: As part of a more long-term oriented planning approach, representatives from corporate logistics, urban planning and economic development are supposed to co-operate in the preservation of freight related locations and land uses for future logistics purposes, even for those that are currently out of operation. This includes the development of a particular action plan for inner-city terminals, e.g. by maintaining railway operations and reserving intermodal nodes for logistics land use rather than to develop the sites for retail, office or other commercial land uses. Such activity is supposed to ensure a major participation of railway and barge modes in future Berlin commercial freight transport. The land use element has been an important feature of the updated concept, particularly because it is not confined to further develop the three freight centres; it is also concerned about inner-city locations in Berlin (e.g. sites formerly used by industry) and to assess their potential for future freight related dedications. As a consequence of the massive post-1989 de-industrialization, there is a significant supply of lots that provides land for logistics, particularly inner-city alternatives to the suburban freight centres. As the empirical findings of this study reveal, a certain portion of logistics firms is attached to the spatio-temporal advantages offered by urban core locations, compared to the more remote suburban freight centres.

Berlin – A Creative Environment for Logistics and Freight Distribution?

Logistics and freight distribution are not only part of transport and urban development plans in either Berlin or Brandenburg, but also included in strategies of regional economic policy. Consequently, it is not only intended to improve the region's performance in transport and logistics, yet also to develop the sector economically. According to self-assessment, Berlin-Brandenburg is already underway to becoming a "centre of competence" for transport technology, management and logistics. A broad range of research institutions, enterprises and public decision makers are working on new transport and logistics projects. In 2006, representatives from both States have launched the "Berlin-Brandenburg Logistics Network" initiative in order to improve the region's freight related performance. This is considered a consequence of corporate networking and particularly of a better international marketing of the region as a "logistics region". Thus the network aims at making customers in Eastern Europe more familiar with the region's competitive advantages. By the way, the intensified marketing of the logistics network may contribute to finally completing the development of the three integrated freight centres. Logistics is also the subject of one out of 16 specialized branch clusters of the State of Brandenburg, since the year 2005 the basis of its regional economic development policy. This applies particularly to the disproportionately high share of commercial development on the Berlin urban fringe, which belongs to the territory of the State of Brandenburg. However, the allegedly high degree of regional activity supersedes the fact that there has been only little implementation in practice until now.

Expectations for innovative logistics activities in the Berlin area were already high a decade ago. During the late 1990s, a broader innovation project named "City of Information Technology" (IT) had been launched in Berlin, initiated by public and private partners which aims at improving the particular IT-competence of the region in several fields of action. In this context, the so-called "City of Logistics" IT-projects had been developed, e.g. in traffic and facility management or in health services. In this way Berlin was to be developed into a logistics model city. Projects such as the "Virtual freight centre" or Hospital logistics had been considered a further indication of the fact that the Berlin participants have a lead over other cities. However, after a few years of action, the project had been abandoned without further notice to the public. The project did not promise to achieve its desired output, primarily due to the lack of implementation, an issue that might have been overlooked once the basic framework conditions of the concept had been investigated. The same happened to activities that had been undertaken in the context of "Transport and Logistics 2000" on behalf of the State Chancellery of the State of Brandenburg and the Senate Chancellery of the Berlin Senate. Despite high ambitions, these activities turned out to be without any practical implications for organizing freight distribution and logistics.

The premise that underlined such research and collaboration efforts was that Berlin is a seedbed of innovative activity in transport research and action. This property is considered to be an outcome of the dense clustering of related institutions in the city, both regarding university research, e.g. at the Technical University of Berlin, and a certain number of non-university research institutions, also private activity carried out by firms such as *Siemens, Deutsche Telekom* or *Daimler*. More recently, intermediate institutions such as the "Research and Application Network Berlin-Brandenburg" (FAV) have been established as well in order to strengthen the links between the public and the private by acquiring EU or Federal government funding and launching research and implementation projects. Whereas such concepts are mainly confined to transport and communications technologies, the link between mobility in more generic terms and urban development is not yet established.

In light of the somewhat disillusioning review of the freight and logistics concept, the question arises whether regional economic and technology policy may in effect contribute to solving the transport dilemma as well. The context of both the regional economic policy and the (freight) transport concept is characterized by structural-economic and transport related goals. At first glance, they appear almost equivalent. However, the practical efforts of research and technology policy in Berlin-Brandenburg pursue first a predominant competition and economy related objective: Berlin is to be established as a business centre for traffic engineering and technology. The promised reduction of motorized traffic and its environmental burden is not supported by empirical evidence yet. Both targets – structural policy and traffic control – are not necessarily congruent. The participants operate in separate environments: both are still far away from a successful implementation of their politics as a study on the state of a railway-based industry cluster had revealed (Dybe and Kujath 2000). Coming closer towards innovation requires linking the two action environments or milieux – the regional-economic and the planning one – more strongly, taking into account each strategies and mutually integrate research and planning goals. This would make the entire concept more credible and might contribute to re-thinking traditional ways of decision making particularly in the transport planning community. There is some evidence that the conditions exist for establishing a collective learning process and effective policy arena in Berlin-Brandenburg – including the case of freight traffic and logistics.

Policy and Planning in The Bay Area/The Central Valley

Spatial planning in the Bay Area has been under particular pressure for decades to accommodate and manage the extraordinary growth of population and employment that characterizes the region for a considerable part of the twentienth century. Besides the unusually high growth rates, specific framework conditions have to be taken into account making the Bay Area policy and planning regime distinct from respective patterns in Berlin-Brandenburg. First, a rapid urbanization has occurred over the last century, starting in the core Bay Area and moving towards increasingly remote places. Initially, the resulting urban formation depended upon the conditions of the natural environment (physical, topographic) and thus limited the opportunities for further growth. Second, commerce and industry are given high priority in politics, making corporations quite powerful and effective in pursuing their goals. In this regard, private development and land speculation have always been important drivers of land use. Third, a strong environmental opposition against further development has emerged over the last 15 to 20 years, particularly in areas that are already urbanized, both resulting in the protection of open space in urban core areas and in pushing development toward places far beyond.

Spatial planning has been developing according to this complex and contested arena of urbanization, growth, and land and infrastructure provision. Barbour (2002) distinguished three related phases in the development of planning practice in California: Urban and regional planning emerged in the "Progressive Era" in the early twentieth century. After the establishment of the "home rule", any municipality could develop in a more or less autonomous way. This is considered being one the major determinants of suburbanization. A second wave of political reforms occurred after World War II approaching suburbanization and focussing on the strategy of "vertical regionalism" besides local autonomy. Also, a broad range of activities in urban renewal had been launched, without having an important impact on planning practice (Frieden and Sagalyn 1989). Since the 1980s, a third wave of planning innovations had emerged putting more emphasis on regional integration and "growth management" (Porter 1997). Due to the lack of an appropriate formal institutional setting, collaborative planning (Healey 1997) is increasingly being pursued since then. More recently, the accelerated sub- and ex-urbanization have called for an intensified search for urban integration, delivered by neo-traditional neighbourhoods, transit-oriented development, and is also subsumed under labels such as "Smart Growth" (Calthorpe 1993) or "Metropolitics" (Orfield 1998), in terms of regional development.

However, the movement of commercial and industrial land uses towards the urban periphery and beyond had been fostered by the conditions of land rents and real estate markets, also by the massive improvement of regional accessibility provided by the freeway-system rather than by zoning policies or planning regulation. The supply of infrastructure helped create the traditional American "vernacular" landscape, often the starting point of commercial strip development or sub- and exurban growth in general. Urban planning had attempted to delimit these outward developments, particularly by submitting growth boundaries, less rigid zoning ordinances or, increasingly, by "infill-developments" in the context of smart growth-policies. Suburbanization, sprawl and related strategies and action plans are also discussed and practised in the San Francisco Bay Area for a considerable time, which both relates to the high growth pressure on one hand, and the political demand for urban design and landscape preservation on the other hand. In response to these trends the "Bay Vision 2020" had been set up in 1990: an alliance that advocated for improved regional planning and co-operation, which then aimed at the institutional reform of regional planning. However, such plans were rejected by a referendum.

The transport sector is in the heart of this controversy, since it is both a cause and a consequence of suburbanization. With increasing traffic volumes, associated problems such as congestion and also air pollution called for action. This applied particularly for places in the Central Valley, e.g. the Sacramento area, which were more critical in this respect than the meteorologically less problematic Bay Area was. However, the related policies were mainly confined to passenger transport, such as public transit, traffic management etc.; freight transport and logistics had been out of consideration for long. This changed in the year 2002, once a team of consultants was engaged to conduct the "Bay Area Goods Movement Study". This was the first time that a conceptual basis had been laid in the San Francisco Bay Area for the management of freight transport and logistics in an urban and regional context (Cambridge Systematics Inc. et al. 2003). As the concept that was published by the end of 2004 reveals, the location and land use dimension of logistics plays a major role for future challenges and possible strategies.

The Bay Area Regional Goods Movement Study – and Policy

With the passage of the Intermodal Surface Transportation Efficiency Act (ISTEA) by the U.S. House of Representatives in 1991, all states and metropolitan planning organizations were, for the first time, required to consider goods movement in the development of their regional transport plan. In response to this requirement, the Bay Area Metropolitan Planning Organization, namely MTC (Metropolitan Transportation Comission), convened a goods movement roundtable in 1992. Corporate representatives from trucking, maritime, rail and air cargo businesses were to solicit ideas on how MTC could consider goods movement issues in planning and funding decisions. Later on, MTC established a Freight Advisory Council (FAC) that supported the commission in identifying problems and discussing possible solutions.

With the long-range 2001 Regional Transportation Plan (RTP) ahead, MTC included the issue of goods movement in its range of activities and commissioned a study in early 2003. The goals of the study were as follows: first, to identify the economic significance of goods movement in the Bay Area on one hand and to make decision-makers aware of the implications of policy decisions that affect goods movement on the other hand; second, to prepare appropriate investment strategies and policies for making regional goods movement more efficient and also to incorporate goods movement in MTC's long-range Regional Transportation Plan.

The study was carried out in two phases. Phase 1 was dedicated to compiling data bases, to evaluate the impact of logistics on the economy and to assess land use issues in the context of freight and logistics. Phase 2 was devoted to address policy issues, such as air quality, and also general implications for transportation planning. The study consisted of several sub-reports, devoting separate volumes to economics, logistics, commodity flows and infrastructure. Two reports have been issued regarding the land use problem, analysing the regional building base, problems and bottlenecks of land supply, and potential solutions.

Among the associated problems, congestion on the frequently used freight corridors such as the I-880 along the East Bay and the I-580 into the Central Valley, are of major concern for the planning authorities. This does not only apply for the corridors, but also for large hubs in the regional freight system, such as the Port of Oakland in particular and the Oakland area in general. Oakland hosts air-freight facilities at Oakland International Airport and also rail facilities of one of the region's two Class 1 Railroads (Burlington Northern and Santa Fe). Since about 80 per cent of the commodities shipped in the region are being carried by truck, both arterial capacity and air quality problems appear to be most urgent.

However, in terms of locational dynamics, the land use issue also represents a major challenge for policy, planning and for corporate management. Hence it is covered extensively by the goods movement study.

Goods-movement-oriented businesses need access to reasonably priced land, in reasonable proximity to customers, where they can conduct their activities without undue conflicts with neighbouring land uses. This is becoming increasingly difficult in the Bay Area. (MTC 2004, 11)

The case that freight businesses are pushed out of the core Bay Area due to high land costs and the declining availability of sites that fit for transport intensive land uses (already discussed earlier in more detail), is generally criticized. This criticism is based on the assumption that the spatial diffusion of distribution centres particularly towards the urban periphery stretches average distances travelled between the place of freight handling and the customers served. Although "... it is not clear how significant these impacts will be and whether the alternatives are more desirable" (Cambridge Systematics 2004, Task 9 Technical Memorandum, 6), it appears plausible that the locational shift towards the periphery contributes to an increasing traffic participation of freight vehicles. As long as the bulk of freight generating activity takes place at the urban core, such locations are considered more efficient than more remote places.

However, in addition to the pressure of the land market already analysed in the context of the Bay Area case study (see Chapter 5), there is another serious problem emerging that contradicts any policy aiming at providing affordable space for trucking and warehousing firms: Smart Growth policies are mainly focussing on infill-developments in the urban core, which often means to convert vacant lots or low density commercial land uses (e.g. warehousing) into retail or, even more often, housing space. Despite the region's spatial mismatch that is an outcome of an uneven distribution of jobs and housing and thus delivers some critical evidence on the current state of affairs, Smart Growth may not turn out to be a wise policy, judging from the logistics and freight distribution perspective. "It is important that these goods-movements locations be preserved in order to preserve the viability of the freight industry and the economic vitality of the region." (MTC 2004, 11) The more light industrial and warehousing space is being converted, the harder it may become for freight and logistics companies to efficiently serve an area. It may then be more likely as well that commercial land uses dedicated to the handling of freight are facing a comparable challenge than urban port and (later) waterfront areas did before.

It is the goal of the related planning guidelines of the regional Goods Movement Study to avoid such effects and to ensure the integration of freight and logistics needs into regional planning on a regular basis. Therefore, a selection of principles has been developed by the study, to be implemented by MTC and local planning authorities:

1. "In locations that support critical goods-movement needs of the central Bay Area, community benefits must be achieved through the application of best practices in off-site impact mitigation and better business practices, while

still preserving central location options for the goods-movement-oriented businesses."

2. "Some suburban locations must accommodate the region's growing needs for warehouse and regional distribution facilities. These facilities will need to be integrated with current land uses without creating major auto/truck/rail conflicts. This smarter suburban development can be accomplished through new approaches to site layout and street design as well as consideration of targeted locations for key perimeter goods-movement facilities in "freight villages" to reduce conflicts and provide greater efficiency."

3. "In consideration of jobs-housing balance, the "jobs" side of the equation must achieve its own balance in terms of diversity of job opportunities for residents with the widest range of skill levels and training. Good-paying jobs at the lower end of the skill range must be preserved and land-use policies and transportation investments should be supportive of this objective."

4. "The trade gateways of the Bay Area — the seaports and airports — represent significant regional assets. Trade is the fastest-growing component of the regional economy and increasing globalization of the world economy portends increasing demands on our gateway facilities. MTC has a particular role to play in ensuring that these facilities remain functional and economically viable. Yet one of the biggest constraints facing these facilities in the future will be the lack of suitable land for supporting businesses and seaport/airport-serving land uses. Regional strategies and incentive programs need to be developed that acknowledge the special needs of communities that house these facilities, so that they will be encouraged to preserve these critical supporting land uses." (MTC 2004, 19).

The strategies and measures compiled by the regional Goods Movement Study indicate that the associated problems can by far not be solved with policies based on the traditional understanding of infrastructure planning and supply. The issue is contradictory and strategies are contested, since they do not serve functionality and sustainability concerns at a time. Also, planning is challenged to appreciate its values in quite a differentiated way, depending on whether transport, land use, environmental or neighbourhood issues are to be solved.

Future Avenues for Policy and Planning

Given the unsustainable character of the freight transport system, particularly in the cases of road and air traffic (OECD 2003), policy and planning will become increasingly important. Yet there is a certain lack of success stories that may pave the way for political intervention. The findings of the two case studies, as different the two regions appear in terms of their planning conditions, are both far from delivering recipes. Due to the complexity of the subject and the different framework conditions, the window of opportunity to manage logistics from a public policy perspective appears to be quite narrow (Banister 2002, 126ff.). Moreover, logistics and freight distribution are not only considered an outcome of a complex structure, but subject to competition, market power and core corporate interests. With direct incentives for firms to minimize transport costs largely missing, any attempt of public intervention turns out to be a weak instrument in a contested arena. This is particularly true since policy and planning have for a long time been based on infrastructure provision, being mostly the responsibility of the public sector. More recently, we observe the emerging role of multinational enterprises that are able to establish their own dedicated infrastructure network (see Graham and Marvin 2001; with respect to port terminal operations: Slack and Frémont 2005). Thus our traditional understanding of infrastructure policy, focusing on publicly provided facilities for general use and wealth, is being challenged by the new kind of meta-privatization of a former public good. It needs to be further investigated how this paradigm shift in infrastructure policy may shape transport policy and intervention in general.

With an eye to the follow-up costs of freight transport, problem-oriented research has long attempted to find solutions for the future, such as inter-modality in long-distance transport or innovation in urban logistics. But the difficulties involved show that the problems are fundamental and systemic in character: The classical method of regulation via infrastructure offers is increasingly unproductive; the companies have a much greater choice of carriers, terminals, space and time optimisations. Judging from the public policy perspective, logistics and freight distribution need to be better integrated into local and regional transport planning. Despite the often claimed significance of freight, many places still concentrate on passenger transport and failed to cope with logistics. As a part of strategy development, building an information base at the regional level is also required. Until now, even the extensively covered field of freight policy in Berlin is still lacking an adequate database. Most of the reviewed planning and policy activities are based on few calculations and estimations. Hence, it is essential to broaden the knowledge base for future freight planning and policy activities (Woudsma 2001). This is particularly true for a more sophisticated, qualitative knowledge about corporate cultures and decision making processes. Substantial difficulties in acquiring the desired data also have to be taken into account, since much logistics information includes confidential corporate data on materials and traffic flow that are hard to retrieve or difficult to simulate.

Secondly, policy and planning activities should concentrate on particular spatial "contexts" of urban and regional freight transport. Given the lesson of construction logistics, such a framework could be supportive in order to make planning work more effectively and more accepted by firms. In the case of Berlin it is, first, the city centre with its extraordinary extent of size, mixed use, housing and offices. On the one hand, it is characterized by an increase of service-related mobility for enterprises and households, which is less remarkable and disturbing than the usual lorry traffic, but occurs in a substantial magnitude. On the other hand, as the local delivery improvements have demonstrated, the core city offers a certain environment (milieu) for regulation. It turns out that market dynamics allow a certain adaption of logistics to the built environment. Hence this field of policy can be actively occupied by corporate and public participants. The second major area of concern is the urban fringe, where an increasing part of the regional logistics facilities have moved to and will continue to do so in the near future. These new logistic nodes with motorway, port, rail yard, and airport access, deserve more attention in terms of urban planning, land use and traffic management. Unlike city centres, those places exemplify the adaptation of space to the requirements of logistics and distribution. This is both true for planned IFC and more scattered freight locations, yet it does not mean the absence of any policy option at all. If transport access in urban regions is becoming a scarce resource, such places offer certain advantages and may experience further growth. Thus they should be acknowledged as subject to planning and design considerations, since they appear economically powerful, but are often nothing more than poorly designed fragments of the metropolitan fringe.

Thirdly, as a consequence of the limitations in finding regional answers to generic structures (as it often occurs in logistics), it is necessary to develop economic incentives and disincentives to enterprises in order to bring individual and common ratio more in line. General fiscal or economic instruments are very suitable for this. The application of such instruments has recently been discussed with respect to the ecological tax reform and to the introduction of a heavy-duty vehicle toll on motorways, which had been introduced in Germany after many delays in 2004 and is now assessed as being successful. However, it has to be recognized that fiscal policies cannot be implemented mainly locally, e.g. in cities; this may lead to substantial competitive distortions among corporate participants or cities and regions. Hence, national and European politics are required in order to support regional activities. As embedded in the broader context of the "New Deal for Transport", the British concept "Sustainable Distribution" (DETR 1999) has once given an example for such a supportive policy framework – even if it did not keep what it initially promised to do. Table 6.2 includes the broad range of strategies and measures that could be undertaken by policy and planning in order to make the system of logistics and freight distribution more sustainable and more acceptable at the local and regional level.

Finally, the issue of energy is certainly changing the environment in which global and regional freight distribution evolves. On one hand, this system is particularly vulnerable to the rapid increase of the oil price, namely due to a high reliance on trucking and air-freight to support time-based distribution. On the other hand, steadily dropping costs may have been a major constraint for policy and planning in the past, particularly the assumption of low energy costs was a major precondition for the recent re-structuring of the logistics industry and the re-organization of logistics networks, resulting in a significant rise of the distribution areas and the associated transport volumes. The long term trend of rising oil prices, the convergence of supply, distribution and refining constraints will indeed have

Table 6.2   Fields of action and measures for local and regional freight policies

Spatial planning, location	Infrastructure, networks	Legal/fiscal policies	Information & communication
Strategic integration of land use and freight transport	Providing intermodal accessibility of commercial areas	Enacting environmental zones, congestion charging	Assessing and monitoring freight flows, particularly in areas of high density
Protecting existing logistics sites, rail yards, inner city ports from re-development	Demanding rail access for newly developed sites	Offering flexible delivery windows in retail and commerce	Promoting local business and planning dialogues
Establishing integrated freight centres for long-haul	Converting light rail systems for goods movement purposes	Providing dedicated delivery zones, lorry ramps etc.	Promoting national freight transport plans and initiatives
Providing inner-city distribution centres for local consolidation and delivery	Establishing new infrastructure (e.g. pipelines, letter shoot)	Installing dedicated lorry lanes on major inner city arterials	Building institutional capacity and responsibility for logistics and freight distribution (e.g. in retail and industry)
		Introducing motorway lorry fees at the national level	Integrating logistics and freight issues in corporate environmental management and corporate social responsibility

Source: own after Flämig 2007

a major impact on the economic sustainability of the transport industry and force substantial adjustments, yet it will also support all those calls for sustainability that are currently not backed by the right signals from the market. It is indicative that, as a consequence of increasing productivity pressures placed on existing transport capacities, rising sensitivity for costs will go along with higher environmental awareness. Whereas entropic forces had been unleashed in freight distribution until recently, the issue of environmental sustainability will become more relevant in the near future, if it is not becoming an issue right now.

Chapter 7

Stability and Change:
Locational Dynamics of Logistics
in an Urban Context

The City as a Terminal: Main Research Findings

The main statement of this book is that modern logistics is shaping urban development and urban land use as a consequence of new supply chain organization and network design. Thus structural changes are creating new geographies of distribution, as an outcome of supply chain management and logistics network design and in response to a changing macro-economic framework. This transformation of urban places includes, first, the re-development of warehousing districts, inner-city rail yards and freight consolidation facilities, in favour of more valuable and competitive land uses, such as housing, retail or business services; this also applies to the increasingly popular conversion of city ports into urban waterfronts. As a consequence, second, facilities that host logistics services, e.g. the storage, consolidation and distribution of consignments, are going to be re-located toward strategic places within and increasingly beyond the limits of urbanized areas.

In order to analyse these questions empirically, the scientific inquiry in the two case study regions was organized step-by-step, based on five major research hypotheses as follows.

1. Location decisions of logistics firms in urban regions are particularly made with respect to the supply of land, transport access and the configuration of distribution areas (highlighting proximity to customers); this combination of location factors has already been at work in the case of the traditional city, yet currently materializes almost exclusively in the surrounding areas of large urban regions.

2. According to the resulting requirements, distribution and logistics businesses are playing a major role in processes of commercial suburbanization, in particular due to their relatively extraordinary demand for space and their sensitivity against conflicts in urban neighbourhoods. In this respect, they represent the case of geographical industrialization.

3. Logistics firms also represent "pionieers" of urban expansion: They stimulate subsequent commercial developments, due to advantages in

location and accessibility. In this respect, they do not only support the process of suburbanization, but also contribute to post-suburbanization.

4. Logistical rationalization and urban expansion in the U.S. can be considered also exemplary in terms of the assessment of future urban development in Europe. This applies particularly for tendencies of ex-urbanization and also for the development, implementation and application of new technologies (e.g. warehouse management systems, efficient consumer response, e-commerce) which appears quite advanced compared to related developments in Western Europe.

5. The assessment of spatial structures and spatial development, e.g. related to the conflict between dispersion and density, is increasingly complex and differentiated: the more suburbanization or even ex-urbanization are becoming predominant, the higher will be the need for developing more appropriate reference points. In this respect, the poly-centric urban region might be the guideline for assessing urban and metropolitan futures.

Towards a Typology of Locational Choice

How do these hypotheses appear in light of the empirical findings of the study? Distribution firms significantly contribute to the process of commercial suburbanization. This has to do with specific locational requirements (transport access, demand for space, proximity to customers) and is, in more general terms, also related to the increasingly networked, service oriented economy. Places of such logistics concentration create regional distribution complexes, based on infrastructure and corporate investment. Site selection practised by such firms is the result of a specific assessment of the pros and cons of agglomeration, rather than being based on a general law of factor combination ("economics remains unchanged, while economies are changing".) Such assessment is closely connected with the competitive profile of the corporation and its either orientation towards core city or region.

i.   The empirical findings reveal that in the distribution business, the inclination to move to suburban places is much more prevalent than the motifs are to remain in the urban core. This is particularly true for major retail chains and their placement of distribution centres, of which only one single firm in Berlin-Brandenburg still locates in the core city area. This also applies to freight forwarding firms which are operating retail distribution chains (except a small percentage only). By and large this locational behaviour appears to be generic. Among the reasons for chosing a suburban location, transport and land related issues appear to be predominant; besides neighbourhood conflicts, the size and functionality of lots are most oftenly referred to as factors causing locational shifts.

The suburban location helps realizing economies of scale and typically reproduces the change from the local to the regional level in goods distribution and supply. This large-scale structure of supply and distribution is primarily made possible by maintaining comprehensive control of the logistics chain as it is the case e.g. in grocery retail distribution, which is directed from the DC into owner-operated outlets. Freight forwarders that are commissioned to deliver consignments to their customers' location are more strongly depending on the input of the receivers and thus spatially less flexible. However, they can compensate for by receiving access to the outlets (e.g. at night time) or by the operation of goods sluices. Thus they may optimize their temporal goods flow, without the need to locate in spatial proximity to their delivery points. Such corporations are assigned to the type of "cost minimizer with almost full control over the logistics chain".

ii.   A smaller, heterogeneous group of firms represent the type of "flexible service provider in high competition". These firms rely on close proximity to the distribution area, particularly due to the critical timing of service operations. Among those firms are parcel services and freight forwarders. Depending on the locational setting, these corporations will remain located in the core area of urban regions or on the edge of the city, as close as possible to their customers. Most important in this respect are advantages in terms of time and also temporal constraints in distribution. Older commercial areas are quite appropriate for this land use, since they are characterized as robust against the related traffic and noise emissions generated there. Due to this property, such locations are extremely valued and represent a major resource for core urban distribution land use.

iii.   Customers of intermodal services are often fixed in their locational behaviour, since they need to be close to a port, railyard or airport. In the sample that had been investigated in Berlin-Brandenburg, only one provider of rail freight services and also one of his customers belonged to this group. However, even among this type of firms, there are tendencies of spatial expansion being observed, partly due to the limited provision of space which results in the expansion of the distribution area or even the terminal infrastructure into the hinterland of the hub. In Northern California, the share of rail freight among the corporations investigated is much higher than it is in Berlin-Brandenburg, which corresponds with a different performance of the railway system in the U.S. in general (much higher distances, almost no significant passenger traffic on the network, earlier deregulation of the market).

iv.   Given the locational preferences of the logistics firms, the relative importance of core city and suburbia is changing favouring the latter. In Berlin-Brandenburg, the movement out of the urban area has been more significant than it was in the Bay Area – probably because of rapidly changing framework conditions, whereas spatial development in

Northern California proceeded in a more continuous way. If analysed in the context of a more long-term perspective, things were changing even in the Bay Area, putting an increasing emphasis on suburban areas. Thus distributive services have contributed to commercial suburbanization in both regions in a significant way. However, to some extent, these services remain oriented towards the core city area – depending on the degree of suburbanization of the region in general.

v.   Labour market supply plays a certain role in places where salaries are low and qualification is at the lower level on one hand, and the cost of living is high on the other hand. The San Francisco Bay Area represents one of such places par excellence, where labour markets at the lower and medium levels (e.g. in warehousing and trucking) are under increasing pressure of the high cost of living. Also, several firms mentioned that a reliable workforce is a convincing argument against the move towards suburbia, once employees are not willing either to move or to commute. The workforce criteria is not necessarily linked with qualification per se (even if this is highlighted in location theories in general and applies, empirically speaking, to core urban areas in particular). Trust can thus be considered a certain expression of qualification.

vi.   Internal linkages among the settlers of suburban locations tend to be performed at low levels. There is no evidence suggesting that freight centres or commercial areas represent networks per se. Rather, vertical control and connectivity along the value chain seems to be predominant. Synergy effects and horizontal linkages that are often searched for in regional economic development processes cannot be identified as relevant, neither in California nor in Berlin-Brandenburg. Obviously, a strong vertical link between the logistics provider and the shipping firms appears to be predominant. Such findings confirm the results provided by McCalla et al. (2001) on the low degree of linkage and local infrastructure use by corporations that are located at intermodal terminals. A similar problem occurred at the Port of Oakland: The scarcity of land is to be solved by the shift of port-related land uses and firms toward more remote areas. This also means that even in the case of the port, the degree of local linkages and interconnectedness appears to be limited.

vii. The hypothesis of certain urbanization effects triggered by logistics firms that represent the case of land use "pioneers" could not be confirmed by the findings of the case studies. There are several reasons that may explain this issue: In Berlin-Brandenburg, the process of suburbanization has slowed down and is partly followed by stagnation. Hence the rapid developments of the 1990s had not been continued so far. In California, corporate locational choice is strongly influenced by the regional performance of land markets and housing markets. However, the clear tendency of some of the distribution firms towards mobilizing economies of scale may avoid local networks, rather than to enable them; this also applies to the poor degree of horizontal relationships among the firms. Also, most of the logistics and freight distribution networks are locationally grounded according to the existing settlement structure, rather than creating completely new or advanced spaces.

viii. New technologies such as EDI, RFID, computer-based warehousing management systems and increasingly mobile tools are being used in corporate practice. In contrast, the significance of electronic commerce for most of the firms remained rather limited. Couriers and parcel services were among the few exceptions that benefited from E-commerce deliveries already in the early 2000s. In the two case study regions, only two core on-line distributors could be identified, one from the office equipment business located in Stockton/CA; another one from the pet food business, operating in Fremont/CA. In Germany, many firms were not engaged in this particular field of business for a considerable time, particularly after the first major crisis of the New Economy following the burst of the bubble in March 2000. Even if some corporations now have successfully established supply chains and business operations, completely new locational structures cannot be identified as an outcome of these activities.

ix. The policy and planning related governance of commercial areas dedicated to logistics and freight distribution land use appears to be quite different in the two case study regions. The Berlin-Brandenburg region and thus the two states had been very active in the establishment of the integrated freight centres (IFC). Such policy has been successful until now – though primarily in terms of attracting corporate investment that favoured certain sites within the IFC areas. This applied predominantly for firms operating strong relationships with partners at long distance. In contrast, the traffic management related effect, particularly modal shifts from road to rail and waterway, has not been achieved so far to a significant extent. In California, policy and practice in the field of freight distribution and logistics are still in the beginning, once strategies have just been worked out by the Metropolitan Planning Organization and by the State of California. This time-lag is caused first by the significantly lower degree of planning intervention into corporate affairs that characterizes the U.S.-situation, compared to Europe; also by the strong functional approach of transport planning and infrastructure provision, which is in Europe more related to aspects of acceptability and sustainability of transport. Also, to some extent different from passenger transport, there seems to be no convincing alternative to the current status quo of freight operations on congested arterials. Even if the Berlin-Brandenburg region is considered a centre of competence in transport and technology issues, it is unlikely that the California region might learn from experience and practice of the German counterpart.

x.   However, the institutional frame for dedicating areas to logistics and freight
       distribution land use and for attracting firms is changing dynamically, in
       Germany as well as already in the U.S. This is indicated by the increasing
       activity of international developers on the emerging market of logistics
       real estate. This had already been observed in the Anglo-American
       context for about decades: and since the late 1990s it is also apparent
       on the European Continent. The emergence of real estate developers is
       changing the pattern of political regulation significantly and also brings a
       noticeable increase in power to private players. In comparison, municipal
       institutions cannot cope with this development – not at least because of
       their diverging interests of both economic development and reducing
       traffic, just to name one of the main conflicts that characterize the public
       arena.

       The pressure of such accelerated processes of accumulation and
       capitalization of places compares to the rapidly changing framework of
       institutions and planning regimes. A common goal is to meet the rising
       demand for land dedicated to logistics operations (Hesse 2004). However,
       after the fundamental mobilization of goods- and traffic-flows, now
       the location itself is becoming mobilized triggered by the outsourcing,
       contracting and leasing of sites. Infrastructure is supposed to be used for
       a limited time and thus marketed and capitalized.

Locational Dynamics in Spatial and Temporal Contexts

The locational choice made by distribution firms is understood as an outcome of
a multidimensional process, within which complex bundles of factors and criteria
play a role, also individual appreciations of values, trade-offs and institutional
context. In terms of urban development, logistics locations are normally resulting
from the successful adaptation to constraints and restrictions (land rents,
competing land uses, traffic bottlenecks). In contrast, suburban places are being
shaped under the influence of the settling firm bringing about regional distribution
complexes. In the latter case, corporations do no longer adapt to certain external
factors and criteria, but can shape these places, due to the completely different
conditions for operation. In this respect, distribution firms specifically contribute
to those processes that had been coined "industrial dispersal" or "geographical
industrialization" (Storper and Walker 1989).

However, due to this contribution, do logistics firms thus have the potential to
represent "pioneers" in the context of urbanization? The answer to this question
depends upon the point which among the coporate players takes over the role
of the spatial-structural "leader", and who might be characterized the respective
"follower". The leader is predominant in terms of spatial and substantial structuring,
which means that he puts economic, social or ecologic emphasis on certain areas.
The leader is driver of innovation and provides substance, potentials and the

direction of a certain development. In contrast, the follower takes over the role of filling in a certain structure, thus reinforcing the weight of a certain development, without being predominant.

According to a classical perspective, the role of the leader would have been appropriate for the industry (manufacturing, large sites, subcontractors, service providers), whereas distribution firms were being considered taking over the part of the follower. This was mainly due to the almost ubiquitous supply of location factors, particularly accessibility and competitive pressure for the temporal optimization of value chain. This classification could have been characteristic for initial phases of spatial development processes or cycles. However, the role of logistics in the context of value chain and supply chain management has become widely expanded and improved. According to the innovation potential of logistics, it can be expected that logistics is taking over the role of the leader and thus shapes spatial development after having experienced structural change and made respective location decisions. In this context, logistics would attract further investments of manufacturing firms – particularly those industries that need to be close to logistics service and can thus materialize high cost reductions. This mainly applies to the new, comfortable locations at the urban periphery, rather than to places in the old urban core.

Initially distribution firms did not represent drivers of urbanization processes per se. The case of Tracy/California demonstrates that this property was more related to cheap land and low rents on the housing market that are attracting distribution investments. However, the resulting regional distribution complexes tend to produce considerable infrastructures that may also become relevant for other firms in the future as well. In the case of further growth of such "flow industries", a subsequent development of the location can become effective. The related progression of such place highly depends upon the framework condition, particularly the availability of sites at the right moment, rents, accessibility, a supportive infrastructure and also regulation in any respect, being positive or negative.

In the case that the spatial behaviour of firms and its outcome is to be further explained, the temporal contexts of each location and the respective framework conditions have to be taken into account. This was demonstrated in the case of the specific phase of political and economic transformation that had occurred in the Berlin-Brandenburg region after 1989/90. Depending on the performance of the particular demand for space and the corresponding supply, the period of early and mid-1990s can be distinguished in three different phases within which the genesis of many commercial areas can be explained:

One strategy within the first phase was directed to attract development primarily on existing commercial areas, as far as the sites were appropriate in terms of size, accessibility and a certain industrial-commercial competence (infrastructure, building permit because of the previous use and dedication, etc.); due to the scarcity of land at that time, many of such initial developments had been placed on the sites where the large state-owned agricultural co-operative associations had existed before.

The other strategy of the initial phase was, running at the same time, to acquire investments according to the principle of "first come, first served". Without having provided any building permits, the strong competition among the mayors of the small municipalities of the New Länder triggered a certain improvisation in order to realize the desired development;

During the third phase after 1996 and 1998 respectively, the Joint State Planning Agency of the States of Berlin and Brandenburg had been established. Since then, any building permit for commercial developments had to be assessed in the context of the newly passed regional planning ordinances. To some extent this had worked well in the case of housing and retail developments. In the case of commercial development, the need for investments and subsequent workplaces was generally considered so high that political intervention and planning regulation did not reach the same degree of implementation. In this phase, however, a majority of freight distribution developments occurred inside and outside existing commercial sites. Also, the year 1995 was the starting point of the development of the first sites within the three integrated freight centres at the periphery of Berlin.

These phases can be basically explained in a place-specific, historical context, yet they are far from being unique. In the context of systematic innovations, they can evolve almost everywhere. The activity of pioneers (in the context of the leader-follower scheme) can prepare structures and situations which may not become regulated yet – since there is no appropriate institutional arrangement, despite individual imaginations of the development potentials of the related area. According to a more systematic perspective of regional development, the role of logistics in the process of urbanization can be summarized as follows:

First, a distinction is to be made between areas of historical poly-centricity or a one-core model. As long as there is a hierarchical order, a certain pressure emerges that pushes land uses towards peripheral areas. This movement continues as long as the pressure in the core is at work, and as long this pressure can be materialized in new suburban developments. Once the hierarchical order does not exist, a more fragmented type of suburbanization will be the case.

Suburbanization as a process emanates from historically emerged pressure on land uses in the core and is going to create a new spatial order on land uses under changing economic and societal framework conditions. As a result of such processes, poly-centricity may emerge, which however remains hierarchical, following a gradient from the core to the periphery. A more chaotic order as an outcome of non-hierarchical networks, or a spatial order that follows the opposite gradient (from the periphery towards the centre), may also occur, yet probably for a limited time only.

Contemporary concepts of the physical development of urban areas have assumed that transport and logistics, as processes of the rationalization of adding value and consumption, act as major drivers of spatial expansion. Transport is not considered being territorially important per se, which the users of traffic connections between source and destination are. The development of advanced logistics in the context of value and production chains leads to the emergence of a new operational level in the entire process of adding value und capitalization. Logistics is likely to take over one of the fundamental characteristics of cities and thus of urban development in general, which is the ability to concentrate. It is thus becoming the core spatial area of source and destination of interactions, in the context of a broader structural economic change.

The impact of logistics turns out to be laminar rather than punctual. This importance is increasing, the higher the number of interactions is becoming. Hence logistics is becoming the functional core of interaction, of pre- and post-production and distribution, and thus is becoming the spatial core as well. However, logistics does not rank first in the hierarchy of the entire value-added process. Hence this sector may not be able to occupy spaces that are more important than those in the suburban area, which are produced by the logistics system itself. A core framework condition for this cycle of impacts is the fact of a well functioning, not entirely fragmented urban region. Logistics works as a space connecting, yet not necessarily hierarchically structured network. It is thus shaping the spatial structure of an urban development more with regard to the urban system, rather than according to a unique development. In such cases only, logistics may play the role as a leader of the spatial development and thus of a "pioneer" in the commercial suburbanization process.

Logistics and Freight Distribution as Subject of Comparative Research

The empirical basis of this investigation was delivered by two case studies which have to be analysed and interpreted in different, specific regional contexts. The Region Berlin-Brandenburg is still characterized by the political and economic transformation, bringing about quite disparate "worlds" as a consequence of the long enduring separation of city and society. The sharp divide of East- and West-Berlin had created contrasting and competing socio-economic and political systems. This overall, systemwide difference was a major determinant of urban development and was also expressed in distinct patterns of spatial development in general and suburbanization in particular (see the case study above in more detail).

Against this historical background, suburbanization in the Berlin-Brandenburg Region and also in the New Länder in general is often characterized as a catch-up process, based on the assumption that it follows a tendency of convergence that makes both parts becoming more similar in the end. The same applies to western (capitalist) suburbanization that is often interpreted as being an outcome of a certain "Americanization" of urban development. However, such models are usually over-simplified and causal, they neglect the specifity of either national or regional contexts. In the case of the former western part of Berlin, suburbanization mostly happened within the urban boundary (the Wall), thus contributing to higher densities and a more urban shape of the city, or outside the region in the Old Länder (former West Germany). The former eastern part of the city had been developed according to a completely different understanding of urban development, which was mainly subject to a political-ideological approach ("a new city for a new society"). Suburbanization was, at least officially, not destined to happen. Large housing estates were the major cases of suburban development, now representing archipelagos of high urban density (not urbanity) in a fragmented, sub-urban setting.

The case study documented in Chapter 4 reveals the specific progression of suburbanization, which is now calming down after having reached a peak in 1998. Further developments and additions to existing and functioning suburban centres are at best consolidating. As a legacy of this historically specific frame of suburbanization, a partly sharp gradient between core city and surrounding areas is prevalent; also, suburbanization is likely to continue within the city limits, both in terms of infill and as an outcome of decreasing densities; finally, the surrounding areas located outside the city of Berlin is characterized by an extremely heterogeneous structure with fragmented portions of settlement scattered around a huge, sparsely populated area. In Berlin-Brandenburg, suburbanization occurs in the context of one large core city with a territorially extensive, yet poorly urbanized surrounding area. Insofar, spatial development in Berlin-Brandenburg is determined by the conditions of the transformation process. This limits the degree to which this particular case might be compared with other regions, both regarding the national and even more the international context. However, since urbanization processes in general are re-connected to broader historical paths of development, the non-transformation related framework conditions of the pre-war era come into play. They include a strong industrial and demographic suburbanization at the turn of the nineteenth century, a historically determined poly-centricity, a temporal rupture of suburban processes in the post-war period.

The fact that the San Francisco Bay Area also represents a prototypical case of poly-centric region is to some extent comparable to Berlin, to some extent not. It appears similar insofar that the process of urbanization was initiated as a process of industrialization; and thus made the rise of Berlin to become the Metropolis of the late nineteenth and early twentieth centuries possible. Different from the Berlin-Brandenburg region with its fragmented suburban areas and its ever changing pattern of growth and decline, the Northern California region is even recently being characterized by a significant growth of population and jobs. Particularly the development of the high-tech industries in the Silicon Valley – since the 1980s – has to be taken into account driving spatial development beyond limits and thus determining the shape of region in the near future, as well as in the more recent past.

Once looking at mere numbers, the Bay Area comprises about almost 18,000 square kilometres and is thus more than three times as big as the Berlin-Brandenburg region is (5,400 km^2). The spatial distribution of a population of almost 7 million people reveals much lower densities compared to Berlin-Brandenburg, where 85 per cent of the population of 4.3 million people live in the city of Berlin. The number of jobs comprises almost 3.3 million in the Bay Area and about 1.4 million in Berlin-Brandenburg. The sharp contrast between city and suburbs is prototypical for the German region, whereas the San Francisco Bay Area is characterized by a mature, poly-centric structure and a high degree of urbanization, now pushing growth portions toward neighbouring places such as the Central Valley.

Both with regard to demographics and economics, the San Francisco Bay Area represents an extraordinary dynamic space. However, further growth is restricted by factors such as topography, particularly the shore-lines and slopes, yet is also due to a traditionally critical attitude of the public against further urban expansion. The S.F. Peninsula and the East Bay Area are the industrialized cores of the region and have been highly urbanized, industrialized and made denser. In contrast, the four northern counties of the Bay Area are less dense and urbanized and more rural. In this regard, they are of suburban character as well.

It is also important to note that the assessment of the suburbanization process in the two case study regions needs to be put into the particular regional context. Of course, the political circumstances of Berlin-Brandenburg were quite unique for more than 40 years, since in West Berlin suburbanization processes were curtailed by a fixed boundary whilst in East Berlin suburbanization was limited to a certain extent, it was at least organized in the state-controlled political economy of East Berlin/East Germany. Also the particular shape of the San Francisco Bay Area delimits and shapes the direction and extent of suburbanization. These retentions do not invalidate the findings of the case studies, but it is important to reflect that, on one hand, such circumstances may not apply in other large conurbations. On the other hand, the development of other major city regions is also affected by regional circumstances such as proximity to international borders, by coastal location in the case of seaports, by the barrier effect of a major river within the urban area or in island states, as it is the case in Singapore. So once discussing the hypothesis of a certain dissociation of logistics and freight distribution from the city, both speaking in more general terms and with regard to the accelerated suburbanisation, the very particular local and regional contexts have to be taken into account carefully.

Options and Limitations of Transatlantic Comparison

The comparative approach of the study has been chosen in order to analyze two different regions against the backdrop of distinct societal and political-economic framework conditions. Judging from this perspective, there is a certain appeal to look at spatial development in North America conducting a comparative analysis with case studies in Europe and in the U.S. – far beyond the matter of fact that spatial developments and settlement structures in the U.S. had become dispersed much earlier and more radical than it has ever been the case in Europe.

Suburbanization represents a generic pattern of urban development and can be studied under supporting framework conditions.

No later than after World War II, the United States of America can be considered a major – if not the most important – innovator and precursor of modernization in the Western world. For a significant time the U.S. have functioned as a blueprint for modernity and thus have often been regarded as an opportunity for looking ahead. At first glance, this is also true for urban and spatial development: The more the spatial structures in Europe are developing into a dispersed direction, the more is this process being coined as "Americanization". However, the term is controversially discussed in social sciences and planning studies, due to its deterministic notion (Meyer 1998; Stiftung Bauhaus Dessau/RWTH Aachen 1995). Also, on one hand, a more or less direct comparison between development patterns in Europe and in the U.S. is limited per se, since the economic, social and cultural starting conditions are quite distinct. On the other hand, the look at U.S. developments allows to anticipate certain contents and their different spatio-temporal context. Does the U.S. offer a kind of laboratory for spatial developments, particularly on the issue of the "unbound metropolis" that grows according to a somehow unleashed capitalism? Given the quite different starting conditions between the two Continents, does the U.S. case tell us something significant with respect to the comparative look at European trends? The answer is not given yet. Some authors (e.g. Hoffmann-Axthelm 1992) have already responded in a positive way. However, at this point, the case is not dealing with direct analogies or parallels in terms of urban development. Research interest is more directed towards the question whether there is a specific surplus in terms of insight into urban dynamics and a potential role of logistics.

On one hand, urban regions in the U.S. do indeed represent prototypical cases of late modern settlement structures that have been developing and are also reinforced in mutual interdependence with transport and logistics innovations. The physical infrastructure and the overall transport conditions and technologies are a major determinant and explanatory factor of spatial development in the U.S. However, in traditional theory, such developments have not been considered being closely associated with transport. Yet they have primarily been explained with the effects generated by economic principles, such as concentration or de-centralization. Despite common framework conditions, this might explain why comparable structures have emerged over time and are still emerging.

On the other hand it appears evident that the amount, timing and speed of implementation of technological innovations that characterizes U.S. developments are of major importance. This applies e.g. for the introduction of the standardized Container as a first step towards the industrialization of goods movement, the shift in logistics organization of large carriers from direct routing to "hub-and-spoke" networks, the introduction of electronic data interchange (EDI) or the invention of Global-Positioning-Systems (GPS). In such cases, the U.S. was trendsetter for both R&D and for implementation. "Follower" economies may not be able to stand apart, will they remain competitive. Hence locational patterns and dynamics of logistics and freight distribution firms in the U.S. and in Europe or Germany basically appear as comparable. This will remain true, even if one takes into account that such generic trends may evolve in a highly diverse spatial setting.

Future Avenues for Research

Logistics deserves more attention in the context of spatial economic networks. In his institutional analysis, Schamp (2000, 203) had called on economic geography not just to analyze how production regions develop, but to pay more attention to the links between production and consumption and its organisation in nodes and logistic systems. The mobility of commodities is inherent to any mode of production and consumption: and it is produced and reproduced in this particular context. So there are good reasons for formulating research avenues related to the geography of distribution.

Issues to be covered by future research particularly relate to the question how the distribution system will emerge in terms of costs, benefits and externalities, and finally how it is regulated. First, it is about production systems, since logistics and distribution are traditionally considered a derived entity (Rodrigue 2006). Contemporary production systems are increasingly characterized by vanishing borders between supply, manufacturing and distribution activity. Logistics firms are taking over parts of both manufacturing and retail activity, and the different components of the chain are becoming widely distributed among actors and places. Managerial and operational power can become concentrated in different hands, either of large manufacturers or large retailers or shipping firms who operate major distribution channels. From the logistics perspective, the variety of opportunities appears much more complicated than the somehow simplified dichotomy of buyer- and supplier-driven commodity chains may suggest. But it is still unclear how far a flow-oriented economy actually represents an independent sector that creates its own functional logics and spatial relations. The tendency towards organisational and spatial separation from the requirements of the shippers in trade and industry is probably not general but selective, focused on specific sections and branches. Outsourcing and the high profile of specialised service providers within the logistics system have blurred the conventional boundaries between manufacturing and marketing, between the logistics of production and of distribution (Visser and Lambooy 2004).

One of the major questions to be answered in future is the role of logistics in the increasingly globally organized production networks. Will logistics and freight distribution continue to be the servant of the production sector, whose performance is more or less derived by the particular demand of manufacturing for timely supply and delivery of resources and components, or is it becoming integrated, or "structural" (Rodrigue 2006)? How might logistics issues change global production networks into globally integrated production and distribution networks, and what part will be playing the major role in this context? In particular, the recently much-discussed global production networks (Coe et al. 2004; Dicken et al. 2001) are closely related to this viewpoint: for the networks observed there cannot be represented without communication and transport or logistics and distribution. In addition to the question of institutional changes, as indicated by the increasing importance of a new kind of service companies, the so-called third- and fourth-party logistics providers, it is also a function of both production organization and a proper transport environment. Only the physical nets enable regions to be integrated into the multi-scalar network of the globalized economy (Hesse and Rodrigue 2006).

In so far as geography and the spatial sciences dealt with the field of traffic and transport in the past their work was strongly shaped by the paradigm of lower – if not completely eliminated – transport costs. In the future, however, energy prices are likely to rise further, leading to higher transport costs, as it already was the case in the year 2007. What impact would this have? As already discussed above, considering costs a major determinant of policy and planning (see Chapter 6), this question is certainly relevant in more generic analytical terms: Would economic activity become concentrated again, would the chains even become re-integrated, and would regional conditions of value creation emerge again? Or would this be offset in the companies' internal calculations by trade-offs between transport and other costs? Location and accessibility, traditional components in cost-based assessments of transport, will see renewed focus. Balances between modes, locations, times and costs will be re-examined to mitigate growing mobility costs with the timely requirements of distribution. A reverse trend in logistics may also take place with customers willing to trade off more time against lower costs.

From the company's perspective the cost dimensions are extremely complex to say the least: They require the analysis not only of transport costs but also of logistics costs and transaction costs. In addition, physical transport generates external costs that can reach very high levels (Infras/IWW 2004). Finally, the cost factor is also related to location and to regions: On balance, do regions such as Berlin-Brandenburg or the Central Valley communities profit from wanting to become transit area and hub, from advertising themselves as "logistic regions"? Obviously not all regions can be equally successful being differently endowed in this respect. Apart from that, such a strategy also has its risks and follow-up costs. Hence, geographical questions such as the location of a certain activity remain dependent upon accessibility in a very physical sense – even though organizational and institutional friction is adding to the pure transport costs. The more transport is becoming critical and a scarce resource, less well functioning and ubiquitous than before (as it is the case in major metropolitan regions and across the transnational transport corridors), the higher will be the emphasis on distribution efficiency, transport time, reliability and, finally, on transport costs.

Finally, if urban places are adapting the property of becoming a "terminal", the question remains what kind of consequence may evolve for the city in general. Whereas this book has investigated two different spatial dimension of logistics and freight distribution, with particular emphasis on suburban areas as the major logistics "organization space" and locational context, it also discusses the generic urban attachment of logistics functions and related processes of dissociation from the city. In this respect, a challenging research avenue is offered for further investigating the role cities may play within the emerging global network of flows and chains. Would the notion of "flows" become so strong that it outperforms the traditional spatial fix (both in terms of solution and in terms of material ground) that urban areas have delivered per se? Does Manuel Castells put it the right way once stating that "the space of flows dominates the space of places"? Such questions would certainly develop in line with the rising interest in the role of cities in global networks that emerged in urban studies recently (cf. Taylor et al. 2007). Under such circumstances, the decisive factors for corporate behaviour and location choce are no longer the concrete place and its amenities, but the rationality of the network and a relational understanding of situation and distance. As a result there is a change in the meaning of situation and location. In addition to "space", "place" and "network", Sheppard (2002) suggested using the term "positionality": this means that the strategic positioning of location in the network space becomes much more important than situation in physical space or in the transport network. Even if such considerations may appear as quite a recent, "modern" issue, the underlying thoughts also easily connect with classical approaches, such as those of spatial interaction (Ullman 1980) or spatial reorganization (Janelle 1969). Centrality, intermediacy and positionality appear as distinct properties of urban places that need to be better understand in the context of the system of flows.

The study of geography offers a significant potential to further explore such relationships and interdependencies between the movement of commodities and information on one hand and the related role that physical places such as cities may play on the other hand. Logistic systems and physical distribution are presently experiencing highly dynamic change, and they make a considerable contribution to the restructuring of systems of spatial relations and locations. This factor has been widely underestimated up to now. However, because of its development dynamics, complexity and structural importance, logistics deserves a legitimate place in research. Judging from the scientific attention it has received as yet, which was highly economic and technical by orientation, this place could appropriately be positioned within geography.

References

ABAG/Association of Bay Area Governments (1997), *Bay Area Futures. Where will we live and work?* (San Francisco: ABAG).

ABAG/Association of Bay Area Governments (1998), *Interdependence. The Changing Dynamic between Cities and Suburbs in the San Francisco Bay Area.* (San Francisco: ABAG).

ABAG/Association of Bay Area Governments (2003), *Census Tracks.* (San Francisco: ABAG).

Abbey, D., Twist, D. and Koonmen, L. (2001), "The need for speed: impact on supply chain real estate", AMB Investment Management, Inc. White Paper. January 2001.

Abernathy, F., J. Dunlop, J. Hammond, and Weil, D. (2000), "Retailing and supply chains in the information age", *Technology in Society* 22:1, 5–31.

Aengevelt Research (1999), *Gewerbegebiet-Report Region Berlin-Brandenburg Nr. 1. Der Markt für Industrie- und Gewerbeflächen im engeren Verflechtungsraum Berlin-Brandenburg.* (Berlin: Aengevelt Immobilien KG).

Alicke, K. (2003), *Planung und Betrieb von Logistiknetzwerken. Unternehmensübergreifendes Supply Chain Management.* (Berlin, Heidelberg: Springer).

Allaert, G. (1999), "The iron Rhine, key issue for cross bordered development". Paper presented at the 1st Scientific Euregional Conferences (SEC), Maastricht November 17–18, 1999. Unpublished Manuscript.

Amin, A. (1999), "An Institutionalist Perspective on Regional Development". *International Journal of Urban and Regional Research* 23:2, 366–78.

Amin, A. and Thrift, N. (1992), "Neo-Marshallian nodes in global networks", *International Journal of Urban and Regional Research* 16:4, 571–87.

Amin, A. and Thrift, N. (2002), *Cities. Reimagining the Urban.* (Cambridge: Polity Press).

Aring, J. (1999), *Suburbia – Postsuburbia – Zwischenstadt.* (Hannover: ARL) = Arbeitsmaterialien 262.

Aring, J. and Herfert, G. (2001), "Neue Muster der Wohnsuburbanisierung", in Brake, K. et al. (eds): *Suburbanisierung in Deutschland – Aktuelle Tendenzen*, pp. 43–56. (Opladen: Leske und Budrich).

Bade, F.-J. and Niebuhr, A. (1999), "Zur Stabilität des räumlichen Strukturwandels", in: *Jahrbuch für Regionalwissenschaften* 19, 131–56.

Bagwell, B. (1982), *Oakland. The Story of a City.* (Oakland: Oakland Heritage Alliance).

Bahrami, B. (2001), "E-Commerce: Implications for Supply chain productivity, carrier competitiveness, and efficient allocation of the external costs of transportation". Paper submitted to The Transportation Research Board, 80th Annual Meeting, Jan. 7–11, 2001. (Washington D.C.: TRB).

Banister, D. (2002), *Transport Planning*. 2nd Edn (London: Spon Press).

Barbour, E. (2002), *Metropolitan Growth Planning in California, 1900–2000*. (San Francisco: Public Policy Institute of California).

Barry, A. and Slater, D. (2001), "The technological economy. Introduction", *Economy and Society* 31:2, 175–93.

Bathelt, H. and Glückler, J. (2002), *Wirtschaftsgeographie. Ökonomische Beziehungen in räumlicher Perspektive*. (Stuttgart: Ulmer).

Baumgarten, H. and Thoms, J. (2002), *Trends und Strategien in der Logistik. Supply Chains im Wandel*. (Berlin: Technische Universität).

Bay Area Economic Forum (2006), *Bay Area Economic Profile 2006*. (San Francisco: BAEF).

Bay Area Economic Forum, Association of Bay Area Governments (2002), *After the Bubble: Sustaining Economic Prosperity*. Bay Area Economic Profile, January 2002. (San Francisco: BAEF).

Bay Area Economic Forum, in Cooperation with Bay Area World Trade Center (2003), *International Trade and The Bay Area Economy*. January 2003. (San Francisco: BAEF).

BBR/Bundesamt für Bauwesen und Raumordnung (2005a), *Raumordnungsbericht 2005*. (Bonn: BBR) = Berichte 21.

BBR/Bundesamt für Bauwesen und Raumordnung (2005b), *Arbeitspapier Raumstrukturtypen. Konzept – Ergebnisse – Anwendungsmöglichkeiten – Perspektiven*. (Bonn: BBR).

Becker, H., Jessen, J. and Sander, R. (1998), *Ohne Leitbild? Städtebau in Deutschland und Europa*. (Stuttgart/Zürich: Karl Krämer).

Behala/Berliner Hafen- und Lagerhaus GmbH (2000), Geschäftsbericht 1999. (Berlin: Behala).

Berliner Bankgesellschaft (2001), *Die Wirtschaftsregion Berlin-Brandenburg zwei Jahre nach dem Regierungsumzug*. Regionalreport. (Berlin: BG).

Bernhardt, C. (1998), *Bauplatz Gross-Berlin. Wohnungsmärkte, Terraingewerbe und Kommunalpolitik im Städtewachstum der Hochindustrialisierung (1871–1918)*. (Berlin: de Gruyter) = Veröffentlichungen der Historischen Kommission zu Berlin 93.

Bertram, H. (2001), "Der Strukturwandel im Speditions- und Transportgewerbe", in Institut für Länderkunde (eds): *Nationalatlas der Bundesrepublik Deutschland*, 9. Verkehr und Kommunikation, 102–03. (Heidelberg, Berlin: Spektrum).

Bertram, H. and Schamp, E. (1989), "Räumliche Wirkungen neuer Produktionskonzepte in der Automobilindustrie", *Geographische Rundschau* 41, 284–90.

Beyer, W. and Birkholz, K. (2003), "Strukturräumliche Entwicklungstrends in Brandenburgs Randregionen". *IRS-aktuell* 41, 4. (Erkner: IRS).

Beyer, W. and Schulz, M. (2001), "Berlin – Suburbanisierung auf Sparflamme?!" in Brake, K. et al. (eds): *Suburbanisierung in Deutschland*, pp. 123–150.

Beyer, W. and Zupp, W. (2002), "Langfristige Bevölkerungsentwicklung Brandenburger Städte bis zum Jahre 2040", *Raumforschung und Raumordnung* 60:2, 89–99.

Bieber, D. (ed.) (2000), *Schnittstellenoptimierung in der Distributionslogistik – Innovative Dienstleistungen in der Wertschöpfungskettte. Dokumentation zur Abschlußkonferenz am 4.5.2000 in Köln.* (Teltow: VDI/VDE).

Binford, H. (1985), *The First Suburbs: Residential Communities on the Boston Periphery 1815–1860.* (Chicago: University of Chicago Press).

Bischoff, B. (2001), *Grundstückspreise in Berlin und Brandenburg. Das Grundeigentum – Spezial*, 13. (Berlin: Grundeigentum-Verlag).

Black, W. (1996), "Sustainable Transport: A U.S.-Perspective", *Journal of Transport Geography* 4:3, 151–59.

Black, W. (2001), "An Unpopular Essay on Transportation", *Journal of Transport Geography* 9:1, 1–11.

BMBau/Bundesministerium für Raumordnung, Bauwesen und Städtebau (1996), *Dezentrale Konzentration – Neue Perspektiven der Siedlungsentwicklung in den Stadtregionen?* (Bonn: BMBau). = Schriftenreihe Forschung 497.

Böhme, R. (1996), *Dezentrale Konzentration logistischer Strukturen. Die systemtheoretische Gestaltung eines makrologistischen Güterverkehrssystems für Berlin-Brandenburg.* (Berlin: TU Berlin) = Schriftenreihe A des Instituts für Straßen- und Schienenverkehr 26.

Bonachic, E. with Hardie, K. (2006), "Wal-Mart and the Logistics Revolution", in N. Lichtenstein (ed.), *Wal-Mart. The Face of Twenty-First-Century Capitalism*, pp. 163–88.

Boustedt, O. (1975), *Grundriß der empirischen Regionalforschung. Teil III: Siedlungsstrukturen.* Taschenbücher zur Raumplanung, 6. (Hannover: ARL).

Bovet, D. and Martha, J. (2000), *Value Nets. Breaking the Supply Chain to Unlock Hidden Profits.* (New York: Wiley).

Bowersox, D., Closs, D. and Stank, T. (2000), Ten mega-trends that will revolutionize supply chain logistics. *Journal of Business Logistics* 21:2, 1–16.

Bowersox, D., Smykay, E., and LaLonde, B. (1968), *Physical Distribution Management. Logistics Problems of the Firm.* (New York, London: MacMillan).

Braudel, F. (1982), *The Wheels of Commerce. Civilization and Capitalism in the 15th 18th Century*, Vol. II. (New York: Harper and Row).

Brechin, G. (1998), *Imperial San Francisco. Urban Power, Earthly Ruin.* (Berkeley, Los Angeles, London: University of California Press).

Brewer, A., Hensher, D. and Button, K. (2001), *Handbook of Logistics and Supply-Chain Management.* (London: Pergamon).

BT Commercial (periodically publ.): *Bay Area Warehouse Report.* (Oakland: BT Commercial). http://www.btcommercial.com/ (accessed on 1 May 2003; accessed on 1 June 2007).

Bukold, S., Deecke, H. and Läpple, D. (1991), *Der Hamburger Hafen und das Regime der Logistik.* (Hamburg: TU Hamburg-Harburg) = Beiträge zur Stadt-, Regional- und Transportforschung 7.

BMVBBW/ Bundesministerium für Verkehr, Bau und Wohnungswesen (1999); *Gütersverkehrsmatrizen für die Bundesverkehrswegeplanung* (Bonn: Unpublished report).

Burdack, J. and Herfert, G. (1998), "Neue Entwicklungen an der Peripherie europäischer Großstädte", *Europa Regional* 6 (1998), 26–44.

Button, K. (1993), *Transport Economics*. 2nd Edn (Cambridge: Edward Elgar).

Cabus, P. and Vanhaverbeke, W. (2003), "The Economics of Rural Areas in the Proximity of Urban Networks: Evidence from Flanders", *Tijdschrift voor Economische en Sociale Geografie* 94:2, 230–45.

Calthorpe, P. (1993), *The Next American Metropolis. Ecology, Community and the American Dream*. (New York: Princeton Architectural Press).

Cambridge Systematics, Inc. with The Tioga Group and Hausrath Economics Group (2003), *Regional Goods Movement Study for the San Francisco Bay Area*. (Cambridge/MA: Camsys).

Capineri, C. and Leinbach, T. (2004), "Transport, e-economy, and globalization", *Transport Reviews* 24: 6, 645–663.

Capineri, C. and Leinbach, T. (2006), "Freight transport, seamlessness, and competitive advantage in the global economy". *European Journal of Transport and Infrastructure Research* 6:1, 23–38.

Castells, M. (1985), "High technology, urban restructuring and the urban-regional process in the United States". Castells, M. (ed.): *High Technology, Space and Society*, pp. 33–40. (Newbury Park: Sage). = Urban Affairs Annual Reviews 28.

Castells, M. (1996), *The Rise of the Network Society*. The Information Age: Economy, Society, and Culture, vol. 1. (Malden/Oxford: Blackwell).

Cervero, R. (1989), *America's Suburban Centers. The Land-Use Transportation Link*. (Boston: Unwin Hyman).

Chandler, A. (1977), *The Visible Hand. The Managerial Revolution in American Business*. (Cambridge/London: Harvard University Press).

Chinitz, B. (1960), *Freight and the Metropolis. The Impact of America's Transport Revolutions on the New York Region*. (Cambridge/MA: Harvard University Press).

Christaller, W. (1933), *Die zentralen Orte in Süddeutschland. Eine ökonomisch-geographische Untersuchung über die Gesetzmäßigkeit der Verbreitung und Entwicklung der Siedlungen mit städtischen Funktionen*. (Darmstadt: Wissenschaftliche Buchgemeinschaft). Reprint 1968.

Christopherson, S. (2001), "Lean retailing in marktliberalen und koordinierten Wirtschaften", in H. Rudolph (ed.), *Aldi oder Arkaden? Unternehmen und Arbeit im europäischen Einzelhandel*, pp. 57–80. (Berlin: Edition Sigma).

Clark, C. (1958), "Transport – maker and breaker of cities". *Town Planning Review* 29:2, 237–50.

Coe, N. and Hess, M. (2005), "The internationalization of retail: implications for supply network restructuring in East Asia and Eastern Europe". *Journal of Economic Geography* 5:4, 449–73.

Coe, N., Hess, M., Yeung, H., Dicken, P. and Henderson, J. (2004), "Globalizing regional development: a global production networks perspective". *Transactions of the Institute of British Geographers* 29:4, 468–84.

Comtois, C. and Rimmer, P. (1997), "Transforming the Asia Pacific's strategic architecture: Transport and communications platforms, corridors and organisations", in Harris, S. and Mack, A. (eds), *Asia – Pacific Security: The Economics-Politics Nexus*, pp. 206–26. (Canberra: Allen and Unwin).

Council of Logistics Management (1986), *What's it All About?* (Oak Brook: Council of Logistics Management).

Cox, A. (1999), "Power, value and supply chain management", *Supply Chain Management: An International Journal* 4:4, 167–75.

Cox, A., J. Sanderson and Watson, G. (2000), *Power Regimes. Mapping the DNA of Business and Supply Chain Relationships*. (London: Earlsgate).

Crang, M. (2002), "Commentary", *Environment and Planning A* 34, 569–74.

Cresswell, T. (2001), "The production of mobilities", *New Formations* 43, 11–25.

Cronon, W. (1991), *Nature's Metropolis. Chicago and the Great West*. (New York: Norton).

Currah, A. (2002), "Behind the web store: the organisational and spatial evolution of multichannel retailing in Toronto", *Environment and Planning A* 34:8, 1411–41.

Daskin, M. and Owen, S. (1999), "Location models in transportation", in Randolph W. Hall (ed.): *Handbook of Transportation Science*, pp. 311–60. (Boston/ Dordrecht/London: Kluwer).

De Ligt, T. and Wever, E. (1998), "European distribution centres: location patterns". *Tijdschrift voor Economische en Sociale Geografie* 89:2, 217–23.

Debernardi, A. and Gualini, E. (1999), "Die Geographie der logistischen Dienstleistungen in der Stadtregion Mailand: Strukturwandel und aktuelle Planungsaufgaben", *Geographische Zeitschrift* 87:3–4, 238–52.

Deecke, H. and Läpple, D. (1996), "German seaports in a period of restructuring", *Tijdschrift voor Economische en Sociale Geografie* 87:4, 332–41.

Deecke, H., Flämig, H. and Hesse, M. (1999), "Zur rechten Zeit am rechten Ort. Vom Anlieferproblem zum theoretischen Konzept der Stadtlogistik", in Friedrichs, J. and Holländer, K. (eds), *Stadtökologische Forschung. Theorien und Anwendungen*, 191–213. (Berlin: Analytica).

DEGI Research/Deutsche Gesellschaft für Immobilienfonds (2006), *Logistikimmobilien. Korridore, Cluster, Märkte*. (Frankfurt/Main: DEGI).

Der Regierende Bürgermeister von Berlin (2000), *Die Berlin-Studie: Strategien für die Stadt*. (Berlin: Regioverlag).

DETR/Department of the Environment, Transport and the Regions (1999), *Sustainable Distribution: A Strategy*. Published 2 March 1999. (London: The Stationery Office).

Dicken, P. (1998), *Global Shift*. 3rd Edn (New York: Guilford).

Dicken, P. (2003), *Global Shift*. 4th Edn (New York: Guilford).

Dicken, P. and Thrift, N. (1992), "The organization of production and the production of organisation: Why business enterprises matter in the study of geographical industrialisation". *Transactions of the Institute of British Geographers* 17:2, 270–91.

Dicken, P., Kelly, P., Olds, K. and Yeung, H. (2001), "Chains and networks, territories and scale: towards a relational framework of analysing the global economy", *Global Networks* 1:2, 89–112.

Douglass, H. (1927), *The Suburban Trend*. (New York: Arno Press).

Drewe, P. and B. Janssen (1998), "What port for the future? From "Mainports" to ports as nodes of logistic networks", in A. Reggiani (ed.) *Accessibility, Trade and Behaviour*, pp. 241–64 (Aldershot: Ashgate).

Dybe, G. and Kujath, H.-J. (2000), *Hoffnungsträger Wirtschaftscluster. Unternehmensnetzwerke und regionale Innovationssysteme: Das Beispiel der deutschen Schienenfahrzeugindustrie.* (Berlin: Edition Sigma).

Easterling, K. (1999), *Organization Space: Landscapes, Highways, and Houses in America.* (Cambridge, London: The MIT-Press).

EDAB/Economic Development Alliance for Business (2006), *East Bay Indicators 2006.* (Oakland: EDAB).

Eisenreich, D. (2001), *Standortwahl und wirtschaftliche Verflechtungen unternehmensorientierter Dienstleistungsbetriebe in der Filderregion. Tendenzen der Abkopplung suburbaner Räume.* (Frankfurt/Main: Peter Lang) = Europäische Hochschulschriften V 2706.

Eno Foundation, Intermodal Association of North America (eds) (1999), *Intermodal Freight Transportation.* 4th Ed. (Washington D.C.: Eno).

Ericksson, T. (2001), "Urban Freight Economics: A New Rail Paradigm For Large Lots". *Transportation Journal* 40:3, 5–15.

Escher, F. (1985), *Berlin und sein Umland. Berlin: Colloquium Verlag* (Berlin: de Gruyter) = Einzelveröffentlichungen der Historischen Kommission zu Berlin 47.

European Commission (1998), *COST 321 – Urban Goods Transport. Final report of the action.* (Brussels, Luxembourg: European Commission).

Fishman, R. (1987), *The Burgeois Utopia. The Rise and the Fall of Suburbia.* (New York: Basic Books).

Flämig, H. (2007), Nachhaltigkeit in der Logistik – Die neuen Herausforderungen an das Management, in *17. Hamburger Logistik-Kolloquium "Nachhaltigkeit in der Logistik"*. Tagungsdokumentation 6 March 2008, pp. 5–12. (Hamburg).

Flick, U. (1991), *Handbuch Qualitative Sozialforschung* (Weinheim: Beltz).

Frey, W. (2003), "Melting Pot Suburbs. A Study of Suburban Diversity", in Katz, B. and Lang, R. (eds), *Redifining Urban and Suburban America*, pp. 155–79. (Washington D.C.: The Brookings Institution).

Frieden, B. and Sagalyn, L. (1989), *Downtown Inc. How America Builds Cities.* (Cambridge, London: MIT Press).

Frigant, V. and Lung, Y. (2002), "Geographical proximity and supplying relationships in modular production". *International Journal of Urban and Regional Research* 26:4, 742–55.

Fulton, W. (1999), "Urban development options for California's central valley". *Land Lines*, September 1999. (Cambridge/Massachusetts: Lincoln Institute for Land Policy).

Garreau, J. (1991), *Edge City. Life on the New Frontier*. (New York: Doubleday).

Gattorna, J. (1990), *The Gower Handbook of Logistics and Distribution Management*. 4th Edn (Aldershot: Gower).

Gereffi, G. and Korzeniewicz, R. (1994), *Commodity Chains and Global Capitalism*. (Westport: Praeger).

Gertler, M. (1992), "Flexibility revisited: districts, nation-states, and the forces of production", *Transactions of the Institute of British Geographers*, 17:2, 259–78.

GL/Gemeinsame Landesplanung Berlin-Brandenburg (1998), *Gemeinsam planen für Berlin und Brandenburg*. Ministerium für Umwelt, Naturschutz und Raumordnung des Landes Brandenburg und von der Senatsverwaltung für Stadtentwicklung, Umweltschutz und Technologie (eds). (Berlin/Potsdam: GL).

GL/Gemeinsame Landesplanung Berlin-Brandenburg (2001), *Erster Flächenbericht zum LEP eV.* (Potsdam: GL).

Glasmeier, A. (1992), "The role of merchant wholesalers in industrial agglomeration formation", *Annals of the Association of American Geographers* 80:3, 394–417.

Glasmeier, A. and Kibler, J. (1996), "Power shift: The rising control of distributors and retailers in the suppy chain for manufactured goods", *Urban Geography* 17:8, 740–757.

Glatzer, R. (1997), *Das Wilhelminische Berlin*. (Berlin: Siedler).

Gordon, P., Kumar, A., Richardson, H. (1988), "Beyond the Journey to Work". *Transportation Research A* 22:6, 419–26.

Gordon, P., Richardson, H. (1997), "Are compact cities a desirable planning goal?" *Journal of the American Planning Association* 63:1, 95–106.

Graham, S. (2000), "Introduction: Cities and Infrastructure Networks". *International Journal of Urban and Regional Research* 24:1, 114–19.

Graham, S. (2001), "Flow city". *DISP* 144, 4–11.

Graham, S. and Marvin, S. (2001), *Splintering Urbanism: Networked Infrastructures, Technological Mobilities and the Urban Condition*. (London, New York: Routledge).

Great Valley Center (2001), *Economic Forecast for California's Central Valley*. (Modesto: Great Valley Center).

Gudehus, T. (2000), *Logistik. Bd. 2 – Netzwerke, Systeme und Lieferketten*. (Berlin, Heidelberg: Springer).

Guy, S. and Henneberry, J. (eds) (2002), *Development and Developers. Perspectives on Property*. Real Estate Issues Series. (Oxford/UK, Malden/MA: Blackwell Science).

Hall, P. (2004), "Mutual specialisation, sea-ports and the geography of automobile imports", *Tijdschrift voor Economische en Sociale Geografie* 95:2, 135–46.

Hall, P., Hesse, M. and Rodrigue, J.-P. (2006), "Re-exploring the interface between transport geography and economic geography". *Environment and Planning* A 38:8, 1401–08.

Handfield, R. and Nichols, E. (1999), *Introduction to Supply Chain Management.* (New Jersey: Prentice Hall).

Hannover Region (2000), *Logistikprofil der Hannover Region.* (Hannover: Kommunalverband Großraum Hannover) = Beiträge zur Regionalen Entwicklung 75.

Hanson, S. (1995), *The Geography of Urban Transportation.* 2nd Edn (New York: Guilford).

Harris, C. and Ullmann, E. (1945), "The nature of cities". *The Annals of The American Academy of Political and Social Science*, 7–17.

Harrison, B. (1997), *Lean and Mean: The Changing Landscape of Corporate Power in the Age of Flexibility.* (New York: Guilford).

Hartwig, N. (2000), *Neue urbane Knoten am Stadtrand? Die Einbindung von Flughäfen in die Zwischenstadt: Frankfurt/Main, Hannover, Leipzig/Halle, München.* (Berlin: Verlag für Wissenschaft und Forschung).

Harvey, D. (1989), *The Condition of Postmodernity.* (Oxford: Blackwell).

Hatton, G. (1990), Designing a Warehouse or distribution centre. Gattorna, J. (ed.), in *The Gower Handbook of Logistics and Distribution Management*, 175–93.

Hausrath Economics with Cambridge Systematics (2003), *Existing Conditions and Trends Regarding Real Estate, Land Use and Community Factors with Implications for Goods Movement Industries. MTC Goods Movement Study Phase II – Task 4 Working Paper.* (Oakland: Hausrath Economics).

Hausrath Economics with Cambridge Systematics (2004), *A Land Use Strategy to Support Regional Goods Movement in the Bay Area. MTC Goods Movement Study Phase II – Task 11 Working Paper.* (Oakland: Hausrath Economics).

Hayden D. (2003), *Building Suburbia. Green Fields and Urban Growth, 1820– 2000.* (New York: Pantheon).

Hayter, R. (1997), *The Dynamics of Industrial Location. The Factory, the Firm, and the Production System.* (Chichester: Wiley).

Hayut, Y. (1981), "Containerization and the load center concept." *Economic Geography* 57:2, 160–75.

Haywood, R. (2001), "Rail-freight growth and the land use planning system", *Town Planning Review* 72:4, 445–67.

Healey, P. (1997), *Collaborative Planning: Shaping Places in Fragmented Societies.* (London: Macmillan).

Heidenreich, G. (1972), *Stadt-Umland-Planung.* (Berlin: Bauakademie der DDR).

Held, D., McGrew, A., Goldblatt, D. and Perraton, J. (1999), *Global Transformations. Politics, Economics and Culture.* (Stanford: Stanford University Press).

Helmke, B. (2005), "Die Wüste lebt". *Logpunkt* 1:3, 12–6.

Henderson, J., Dicken, P., Hess, M., Coe, N. and Yeung, H. (2002), "Global production networks and the analysis of economic development". *Review of International Political Economy* 9:3, 436–64.

Herfert G. (2002), "Disurbanisierung und Reurbanisierung. Polarisierte Raumentwicklung in der ostdeutschen Schrumpfungslandschaft". *Raumforschung und Raumordnung* 60:5–6, 334–44.

Herfert, G. (2003), "Die Peripherie der Stadtregion Berlin – zwischen Boom und Schrumpfung." Unpublished manuscript.

Hesse, M. (1998a), *Wirtschaftsverkehr, Stadtentwicklung und politische Regulierung. Zur Bedeutung des Strukturwandels in der Distributionslogistik für die Stadtplanung.* (Berlin: Deutsches Institut für Urbanistik) = Beiträge zur Stadtforschung 26.

Hesse, M. (1998b), "Städtischer Wirtschaftsverkehr als Gegenstand der kommunalen Planung. Möglichkeiten und Grenzen privat-öffentlicher Kooperation", *Archiv für Kommunalwissenschaften* 37:2, 240–59.

Hesse, M. (1999), "Der Strukturwandel von Warenwirtschaft und Logistik und seine Bedeutung für die Stadtentwicklung", *Geographische Zeitschrift* 87:3–4, 223–37.

Hesse, M. (2002), "Shipping news: The implications of electronic commerce for logistics and freight transport", *Resources, Conservation and Recycling* 36:2, 211–40.

Hesse, M. (2002b), "Location matters". *Access* 21, 22–6.

Hesse, M. (2004), "Land for logistics: Locational dynamics and political regulation of distribution centres and freight agglomerations", *Tijdschrift voor Economische en Sociale Geografie* 95:2, 162–173.

Hesse, M. (2006), "Logistikimmobilien. Von der Mobilität der Waren zur Mobilisierung des Raumes". *DISP* 42:4, No. 167, 43–51.

Hesse, M. and Rodrigue, J.-P. (2004a), "The transport geography of logistics and freight distribution". *Journal of Transport Geography* 12:2, 171–84.

Hesse, M. and Rodrigue, J.-P. (2004b), "Dossier on freight transport and logistics". *Tijdschrift voor Economische en Sociale Geografie* 95:2, 133–35.

Hesse, M. and Rodrigue, J.-P. (2006), "Global production networks and the role of logistics and transportation". *Growth and Change* 32:4, 499–509.

Hesse, M., Schmitz, S. (1998), "Stadtentwicklung im Zeichen von „Auflösung" und Nachhaltigkeit". *Informationen zur Raumentwicklung* 7/8.1998, 435–53.

Hoffmann-Axthelm, D. (1992), "Der Weg zu einer neuen Stadt", *ARCH+* 114/115, 114–16.

Hofmann, W. (1997), "Stadt und Umland", *Informationen zur modernen Stadtgeschichte* (IMS) 2/97, 3–7.

Hofmeister, B. (1990), *Berlin (West). Eine geographische Strukturanalyse der zwölf Westbezirke.* (Gotha: Perthes).

Hollingsworth, J. and Boyer, R. (1997), *Contemporary Capitalism. The Embeddedness of Institutions.* (Cambridge: Cambridge University Press).

Holmes, J. (2000), "Regional Economic Integration in North America", in G. Clark, M. Feldman and M. Gertler (eds) *The Oxford Handbook of Economic Geography*, pp. 649–670. (Oxford: Oxford University Press).

Höltgen, D. (1995), "Güterverkehrszentren und Kombinierter Verkehr — Raumwirksamkeit europäischer Netze", in Barsch, D. and Karrasch, H. (eds): *Verhandlungen des 49. Dt. Geographentags Bochum 1993*, 4, 157–63. (Stuttgart: Franz Steiner).

Holton, R. (2005), *Making Globalization*. (Hampshire: Palgrave/Macmillan).

Hopkins, T. and Wallerstein, I. (1986), "Commodity Chains in the World Economy Prior to 1800", *Review* 10:1, 157–170.

Hopkins, T. and Wallerstein, I. (1994), "Commodity chains: Construct and research", in G. Gereffi and M. Korzeniewicz (eds), *Commodity Chains and Global Capitalism*, pp. 17–20. (Westport: Praeger).

Hoppe, N. and Conzen, F. (2002), *Europäische Distributionsnetzwerke*. (München: Gabler).

Houser, S. (2000), *Inventory of Business and Industrial Parks in Central Valley Cities*. Prepared for the Central California Futures Institute. (Fresno: California State University, Dept. of Economics).

Hoyle, B. (1990), *Port Cities in Context: The Impact of Waterfront Regeneration*, IBG-Transport Geography Study Group. (London: IBG).

Hoyle, B. (1996), *Cityports, Coastal Zones and Regional Change: International Perspectives on Planning and Management*. (Chichester: Wiley).

Hoyle, B. and Knowles, R. (1998), *Modern Transport Geography*. 2nd Edn (London: Wiley).

Hoyle, B. and Pinder, D. (1992), *European Port Cities in Transition*. (London: Bellhaven Press).

Hoyle, B.S. (1996), *Cityports, Coastal Zones and Regional Change: International Perspectives on Planning and Management*. (Chichester: Wiley).

Hoyt, H. (1939), *The structure and Growth of Neighborhoods in American Cities*. (Washington D.C.: Federal Housing Association).

Hudson, R. (2001), *Producing Places*. (London: Guilford).

Hughes, A. (2000), "Retailers, knowledge and changing commodity networks: The case of the cut flower trade", *Geoforum* 31:2, 175–90.

Hummels, D. (2001), Time as a Trade Barrier, www.mgmt.purdue.edu/centers/ ciber/ publications/00–007Hummels2.pdf (accessed on 1 May 2003).

IHK zu Berlin (1995), *Wirtschaftsentwicklung und Raumplanung in der Region Berlin-Brandenburg. Modelle, Tendenzen, Entwicklungserfordernisse*. (Berlin: Regioverlag).

IMF (2001), *World Economic Outlook: The Information Technology Revolution*. (Washington D.C.: IMF).

Infras, IWW (2004), *Externe Kosten des Verkehrs. Aktualisierungsstudie*. (Zürich/ Karlsruhe: Infras/IWW).

INFRAS/IWW (2000), *External Costs of Transport: Accident, Environmental and Congestion Costs in Western Europe*. Report for the International Union of Railways. (Paris, Zurich, Karlsruhe).

IÖR, IRS and Universität Leipzig (eds) (2005), "Mobilität im suburbanen Raum. Neue verkehrliche und raumordnerische Implikationen des räumlichen Strukturwandels". Forschungsbericht für das BMVBW (FE 70.716). Dresden, Erkner and Leipzig, unpublished.

IÖR/Institut für ökologische Raumentwicklung (2002), *Siedlungsstrukturelle Entwicklung im Umland der Agglomerationen.* Forschungsvorhaben im Auftrag des Bundesamtes für Bauwesen und Raumordnung. (Dresden: IÖR).

IVU (2000), Verkehrsentwicklungsplanung Berlin und Brandenburg. Präsentation am 19.12.2000. (Berlin: Unpublished manuscript).

Jackson, J.B. (1992), Truck city, in Wachs, M. and Crawford, M. (eds), *The Car and the City,* pp. 16–24. (Ann Arbor: The University of Michigan Press).

Jackson, J.B. (1994), Truck City, in Jackson, J. B., *A Sense of Place, A Sense of Time*, pp. 171–185 (New Haven/London: Yale University Press).

Jackson, K. (1985), *The Crabgrass Frontier. The Suburbanization of the United States.* (New York, Oxford: Oxford University Press).

Jackson, P., Ward, N. and Russel, P. (2006), Mobilizing the commodity chain concept in the context of food and farming. *Journal of Rural Studies* 22, 129–41.

Janelle, D. (1969), "Spatial reorganization: A model and concept", *Annals of the Association of American Geographers* 59:2, 348–64.

Janelle, D. and M. Beuthe (1997), "Globalization and research issues in transportation", *Journal of Transport Geography* 5:3, 199–206.

JBF Associates et al. (2001), *Rough and Ready Island Development Plan for the Port of Stockton, California.* Final Report Feb 28, 2001. (Stockton: The Port of Stockton; Fort Lauderdale: JBF).

Jones Lang LaSalle (ed.) (2001a), *The Changing Landscape of European Distribution Warehousing*, January. (London: JLL).

JonesLangLaSalle (2001b), *Gewerbegebiet Report Berlin. Der Markt für Büro-, Industrie- und Gewerbeflächen in der Region Berlin.* (Berlin: JLLS).

Jones Lang LaSalle (2006), *Logistikimmobilien-Report Deutschland 2006.* (Frankfurt/M., Hamburg: JLLS).

Kahnert, R. (1998), "Wirtschaftsentwicklung, Sub- und Desurbanisierung". *Informationen zur Raumentwicklung* 7/8.1998, 509–20.

Karsten, M., Usbeck, H. (2001), "Gewerbesuburbanisierung – Die Tertiärisierung der suburbanen Standort", in Brake, K. et al. (eds): *Suburbanisierung in Deutschland*, pp. 71–80.

Katz, B. and Lang, R. (eds) (2003), *Redefining Urban and Suburban America. Evidence from Census 2000, Vol. I.* (Washington D.C.: Brookings).

Kia, M., Shayan, E. and Ghotb, F. (2003), "Positive impact of distribution centres on the environment". *Transport Reviews* 23:1, 105–22.

Kirschbraun, T. and Bomba, T. (2000), The new economy. Effects on corporate real estate strategies. In: Jones Lang LaSalle (ed.): *Property Futures. Occupier Strategies in the New Economy.* (Chicago, London: JLLS).

Kloosterman, R. and Musterd, S. (2001), "The polycentric urban region: Towards a research agenda". *Urban Studies* 38:4, 623–33.

Knowles, R. and Hall, D. (1998), "Transport Deregulation and Privatization", in B. Hoyle and R.D. Knowles (eds) *Modern Transport Geography*, pp. 75–96. (New York: Wiley).

Knox, P. and Agnew, J. (1998), *The Geography of the World Economy*, 3rd Edn (New York: Wiley).

Kostof, S. (1993), *Die Anatomie der Stadt. Geschichte städtischer Strukturen.* (Frankfurt/M., New York: Campus).

Kotkin, J, and Frey, W. (2007), *The Third California. The Golden State's New Frontier.* Brookings Research Brief. (Washington D.C.: The Brookings Institution).

Kotkin, J. (2000), *The New Geography. How the Digital Revolution Is Reshaping the American Landscape.* (New York: Random House).

Kreukels, A. (2003), "Wie verankern sich Häfen im Raum?" *DISP* 154, 26–7.

Kujath, H.-J. (1995), "Zwischen Zentrum und Peripherie. Regionale Transformation und Raumordnungspolitik in Berlin-Brandenburg", in Ipsen, D. (ed.): *Stadt und Region – Stadtregion.* (Frankfurt/M.: Hessische Gesellschaft für Demokratie und Ökologie).

Lakshmanan, T. R. and Anderson, W.P. (2002), "Transportation Infrastructure, Freight Services Sector and Economic Growth", White Paper prepared for The U.S. Department of Transportation, Federal Highway Administration (Boston University: Center for Transportation Studies).

Lambert, D. (2001), "The supply chain management and logistics controversy", in: Brewer, A., Button, K. and Hensher, D. (eds): *Handbook of Logistics and Supply Chain Management*, pp. 99–126. (Amsterdam: Elsevier).

Landesamt für Bauen, Verkehr und Straßenwesen Brandenburg (2002), Daten zur Verkehrsbelastung auf der B 101 in Großbeeren. (Statistical spreadsheet).

Landis, J. and Reilly, M. (2003), *How We Will Grow: Baseline Projections of the Growth of California's Urban Footprint through the Year 2100.* Institute of Urban and Regional Development, University of California, Berkeley. (Berkeley: IURD).

Landtag Brandenburg (2000), Antwort der Landesregierung auf die Grosse Anfrage der PDS-Fraktion 3/1652 vom 29.8.2000. Potsdam.

Lang, R. (2003), *Edgeless Cities. Exploring the elusive Metropolis.* (Washington D.C.: Brookings).

Langlois, R.N. (2003), "The vanishing hand: the changing dynamics of industrial capitalism". *Industrial and Corporate Change* 12:2, 351–85.

Läpple, D. (1995a), "Transport, Logistik und logistische Raum-Zeit-Konfigurationen" in D. Läpple (ed.), *Güterverkehr, Logistik und Umwelt. Analysen und Konzepte zum interregionalen und städtischen Verkehr*, pp. 21–57. (Berlin: Edition Sigma).

Läpple, D. (1995b), "Hafenwirtschaft", in ARL (ed.): *Handwörterbuch der Raumordnung*, pp. 462–71. (Hannover: ARL).

Lempa, S. (1990), *Flächenbedarf und Standortwirkung innovativer Technologie und Logistik.* (Regensburg: Kallmünz). = Münchner Studien zur Sozial- und Wirtschaftsgeographie 36.

Leslie, D. and Reimer, S. (1999), "Spatializing commodity chains", *Progress in Human Geography* 23:3, 401–20.

Lewis, R. (1999), "Running rings around the city: North American industrial suburbs, 1850–1950". In: Harris, R., Larkham, P. (eds): *Changing Suburbs*, 146–67.

Liechtenstein, N. (ed.) (2006), *Wal-Mart. The Face of Twenty-First-Century Capitalism*. (New York, London: The New Press).

LUA/Landesumweltamt Brandenburg (2000), *Großflächiger Einzelhandel im gemeinsamen Planungsraum Berlin-Brandenburg. Analyse der großflächigen Einzelhandelszentren, -betriebe und –einrichtungen ab 5.000 m² Verkaufsfläche.* (Potsdam: LUA). Fachbeiträge des Landesumweltamtes 50.

Marsden, T. and Wrigley, N. (1996), "Retailing, the food system and the regulatory state", in N. Wrigley and Lowe, M. (eds) *Retailing, Consumption and Capital*, pp. 33–47 (University of Southampton: Longman).

Marx, K. (1939/1953), *Grundrisse der Kritik der politischen Ökonomie. (Foundations of the critique of the political economy)*. Berlin: Dietz (Middlesex/ England: Penguin Books).

Masotti, L. and Hadden, J. (eds) (1973), *The Urbanization of the Suburbs.* (Thousand Oaks: Sage). = Urban Affairs Annual Reviews 7.

Matthews, H. and Hendrickson, C. (2003). "The Economic and Environmental Implications of Centralized Stock Keeping". *Journal of Industrial Ecology* 6:2, 71–81.

Matthews, H., Hendrickson, C. and Soh, D. (2001), "Environmental and economic effects of e-commerce: A case study of book publishing and retail logistics". *Transportation Research Record* 1763, 6–12.

Matthiesen, U. (2002), "Fremdes und Eigenes am Metropolen-Rand – Postsozialistische Hybridbildungen in den Verflechtungsmilieus von Berlin und Brandenburg", in Matthiesen, U. (ed.), *An den Rändern der deutschen Hauptstadt*, pp. 327–52.

Matthiesen, U. (ed.) (2002), *An den Rändern der deutschen Hauptstadt*. (Opladen: Leske und Budrich).

Matthiesen, U. and Nuissl, H. (2002), "Phasen der Suburbanisierung seit 1989: Stichpunkte zum Berlin-Brandenburger Verflechtungsprozess", in Matthiesen, U. (ed.), *An den Rändern der deutschen Hauptstadt*, 79–92.

McCalla, R., Slack, B. and Comtois, C. (2001), "Intermodal freight terminals: locality and industrial linkages." *The Canadien Geographer/Le Géographe canadien* 45:3, 404–13.

McCalla, R., Slack, B. and Comtois, C. (2004), "Dealing with globalisation at the regional and local level: the case of contemporary containerization", *The Canadian Geographer/Le Géographe canadien* 48:4, 473–87.

McCray, J. (1998), "North American Free Trade Agreement truck highway corridors. U.S.-Mexican truck rivers of trade", *Transportation Research Record* 1613, 71–8.

McKinnon, A. (1983), "The development of warehousing in England", *Geoforum* 14, 389–99.

McKinnon, A. (1988), "Physical distribution", J.N. Marshall (ed.) *Services and Uneven Development*, pp. 133–59. (Oxford: Oxford University Press).

McKinnon, A. (1998), "Logistical restructuring, freight traffic growth and the environment", in D. Banister (ed.) *Transport Policy and the Environment*, pp. 97–109. (London: Spon Press).

McKinnon, A. and Woodburn, A. (1996), "Logistical restructuring and road traffic growth. An empirical assessment", *Transportation* 23:2, 141–61.

Meijer, M. and Ten Velden, H. (1996), "Spatial aspects of logistics activity", *EUREG* 3/96, 49–54.

Metropolitan Transportation Commission/MTC (2001), *2001 Regional Transportation Plan for the San Francisco Bay Area*. (Oakland: MTC).

Metropolitan Transportation Commission/MTC (2004), *Regional Goods Movement Study*. (Oakland: MTC).

Meyer, H.-D. (1998), Review article on Lüdtke, A., Marssolek, I. et al. (eds): Amerikanisierung. Traum und Alptraum im Deutschland des 20. Jahrhunderts. Stuttgart 1996. *Soziologische Revue* 21, 231–33.

Mokhtarian, P. (2000), "Telecommunications and travel". Millenium paper of the Committee on Telecommunications and Travel Behavior of the Transportation Research Board. (Washington D.C.: TRB).

Motzkus, A. (2002), *Dezentrale Konzentration – Leitbild für eine Region der kurzen Wege?* (Sankt Augustin: Asgard). = Bonner Geographische Arbeiten 107.

Mueller, G. and Laposa, S. (1994), "The path of goods movement". *Real Estate Finance* 11, 42–50.

Muller, P. O. (1989), "The transformation of bedroom suburbia into the outer city: An overview of metropolitan structural change since 1947", in: Kelly, B. (ed.): *Suburbia Re-examined*, pp. 39–44. (New York: Greenwood Press).

Neher, A. (2005), "Internationale Logistikstrategien von Industrieunternehmen im Wandel", in: Neiberger, C. and Bertram, H. (eds) *Waren um die Welt bewegen, Strategien und Standorte im Management globaler Warenketten*, pp. 33–45.

Neiberger, C. (1997), "Die Neuordnung der Logistikkette. Räumliche Auswirkungen unternehmensübergreifender Umstrukturierungsprozesse zwischen Industrie und Handel. Das Beispiel Molkereiprodukte", *Erdkunde* 51:2, 131–42.

Neiberger, C. (1999), "Standortvernetzung durch neue Logistiksysteme: Zur Hersteller-Handels-Integration in der deutschen Nahrungsmittelwirtschaft", *Seminarberichte der Gesellschaft für Regionalforschung* 41, 197–221.

Nelson, R and Winter, S. (1982), *An Evolutionary Theory of Economic Change*. (Cambridge: Harvard University Press).

Notteboom, T. (2004), "Container shipping and ports: An overview", *Review of Network Economics* 3:2, 86–106.

Notteboom, T. and Rodrigue, J.-P. (2005), "Port regionalization: Towards a new phase in port development", *Maritime Policy and Management* 32:3, 297–313.

Notteboom, T. and Winkelmans, W. (2004), *Factual Report on the European Port Sector. Work Package 1: Overall market dynamics and their impact on the Port Sector*. Commissioned by the European Sea Port Organization. (Antwerp: ITTMA).

Nuhn, H. (1999), "Changes in the European gateway system – the case of seaports", *Beiträge zur Regionalen Geographie Europas* 47, 88–102.

Nuhn, H. (2005), "Internationalisierung von Seehäfen. Vom Cityport und Gateway zum Interface globaler Transportketten", in Neiberger, C. and Bertram, H. (eds) *Waren um die Welt bewegen. Strategien und Standorte im Management globaler Warenketten*, pp. 109–24.

Nuhn, H., A. Berthold, C. Neiberger and Stamm, A. (1999), *Auflösung regionaler Produktionsketten und Ansätze zu einer Neuformierung. Fallstudien zur Nahrungsmittelindustrie in Deutschland* (Münster: LIT).

Nuissl, H. (1997), "Räumliche Entwicklungen im „Speckgürtelchen". Planen und Bauen an den Rändern von Berlin", in *RaumPlanung* 77, 109–14.

Nuissl, H. (1999), "Suburbanisierung und kommunale Entwicklungsstrategien an den Rändern der Hauptstadt", in *Archiv für Kommunalwissenschaften* II/99, 237–57.

The New York Times, 17 April 2001, The new-look suburbs: Denser or more far-flung, A1/A14.

O'Kelly, M. (1998), "A geographer's analysis of hub-and-spoke networks", *Journal of Transport Geography* 6:3, 171–86.

Oakley Strategic Economics (2001), "Economic composition of four selected corridors in the Bay Area and the Central Valley" (unpublished; data after Dun & Bradstreet).

Odgen, K.W. (1992), *Urban Goods Movement. A Guide To Policy and Planning* (Aldershot: Ashgate).

OECD (2000), *The Economic and Social Impacts of Electronic Commerce: Preliminary Findings and Research Agenda.* (Paris: OECD).

OECD (2003), *Delivering the Goods. 21ˢᵗ Century Challenges to Urban Goods Transport.* (Paris: OECD).

OECD/ECMT (2001), *The Impact of E-commerce on Transport.* Joint Seminar. (Paris: OECD).

Orfield, M. (1998), *Metropolitics. A Regional Agenda for Community and Stability.* (Washington D.C.: Brookings and Lincoln Institute for Land Policy).

Overwien, P. (2003), *Planungsbezogenes Konfliktmanagement unter Transformationsbedingungen. Ein empirischer Beitrag zur Erklärung von Suburbanisierungstendenzen in der Stadtregion Berlin,* (Berlin: Dietrich Reimer) = Abhandlungen Anthropogeographie, Institut für Geographische Wissenschaften, Freie Universität Berlin 64.

Pawley, M. (1994), "The redundancy of space. Die Redundanz des urbanen Raumes", in: Meurer, B. (ed.): *Die Zukunft des Raums*, pp. 37–57. (Frankfurt/ M., New York: Campus).

Pedersen, P.O. (2000), "Freight transport under globalisation and its impact on Africa", *Journal of Transport Geography* 9:2, 85–99.

Pfannschmidt, M. (1937), "Die Industriesiedlung in Berlin und in der Mark Brandenburg. Ihre Entwicklung vom Absolutismus bis zur Gegenwart und ihre zukünftigen Entwicklungsmöglichkeiten". Herausgegeben von der Akademie für Landesforschung und Reichsplanung. (Stuttgart-Berlin: Kohlhammer).

Pincetl, S. (1999), *Transforming California. A Political History of Land Use and Development*. (Baltimore: The Johns Hopkins University Press).

Piore, M. and Sabel, C. (1984), *The Second Industrial Divide: Possibilities for Prosperity*. (New York: Basic Books).

Pochet, L., Rumley, P.-A. and de Tilière, G. (2000), *Plates-formes logistiques multimodal et multiservices*. Rapports du PNR 41 "Transport et environnement", rapport B 9. (Berne: OFCL/EDMZ).

Port of Oakland (2002), *Port of Oakland Strategic Plan. FSY 2003–2007 Update*. (Oakland: Port of Oakland).

Port of Oakland (2000), *Vision 2000. Maritime Development Program*. (Oakland: Port of Oakland).

Porter, D. (1997), *Managing Growth in America's Communities*. (Washington D.C.: Island Press).

Pred, A. (1977), *City Systems in Advanced Economies*. (London: Hutchinson).

Priebs A (2004), "Vom Stadt-Umland-Gegensatz zur vernetzten Stadtregion", in *Jahrbuch StadtRegion 2003*, pp. 17–42. (Opladen: Leske und Budrich).

Race, B. (2001), *Sprawling beyond the Edges: The Next Wave* (San Francisco: SPUR).

Rae, J.B. (1971): *The Road and the Car in American Life* (Cambridge: MIT-Press).

Raikes, P., Jensen, M. and Ponte, S. (2000), "Global commodity chain analysis and the French filière approach: comparison and critique", *Economy and Society* 29:3, 390–417.

Ralston, B. (2003), "Logistics", in H. Miller and S. Shaw *Geographic Information Systems for Transportation: Principles and Applications*, pp. 380–400 (Oxford: Oxford University Press).

Reulecke, J. (1985), *Geschichte der Urbanisierung in Deutschland*. (Frankfurt/M.: Suhrkamp).

Rich, M. (2001), *The implications of changing U.S. demographics for housing choice and location in cities*. Discussion paper, prepared for the Brookings Institution Center on Urban and Metropolitan Policy. (Washington D.C.: Brookings).

Riehm, U., Petermann, T., Orwat, C., Coenen, C. Revermann, C., Scherz, C. and Wingert, B. (2003), *E-Commerce in Deutschland. Eine kritische Bestandsaufnahme zum elektronischen Handel*. (Berlin: Edition Sigma).

Riemers, C. (1998), "Functional relations in distribution channels and locational patterns of the Dutch wholesale sector", *Geografiska Annaler B* 80, 83–97.

Robeson, J. and Kollat, D. (1985), Channels of Distribution: Structure and Change, in J. Robeson and House, R. (eds): *The Distribution Handbook*, pp. 225–34 (New York, London: The Free Press).

Robins, M. and Strauss-Wieder, A. (2006), *Principles for a U.S. Public Freight Agenda in a Global Economy*. The Brookings Institution Series on Transportation Reform. (Washington D.C.: The Brookings Institution).

Rodrigue, J.-P. (1999), "Globalization and the synchronization of transport terminals", *Journal of Transport Geography* 7:4, 255–61.

Rodrigue, J.-P. (2006), "Challenging the derived transport demand thesis: Geographical issues in freight distribution", *Environment and Planning A* 38:8, 1449–62.

Rodrigue, J.-P. and Hesse, M. (2007), "Globalized freight and logistics – North American perspectives", in Leinbach, T. and Capineri, C. (eds): *Globalized Freight Transport: Intermodality, E-Commerce, Logistics and Sustainability,* pp. 103–134. (Cheltenham: Edward Elgar).

Rodrigue, J.-P. and Slack, B. (2002), "Logistics and national security", in Majumdar, S. (ed.) *Science, Technology, and National Security* (Pittsburgh: Pennsylvania Academy of Science Press).

Rodrigue, J.-P., Slack, and Comtois, C. (2001), "Green logistics", in Brewer, A., Button, K. and Hensher, D. (eds) *The Handbook of Logistics and Supply-Chain Management, Handbooks in Transport* (London: Pergamon/Elsevier).

Romm, J. (1999), *The Internet Economy and Global Warming.* Technical report. (Annandale, VA: Center for Energy and Climate Solutions, Global Environmental and Technology Foundation).

Ryan, S. (1999), "Property values and transportation facilities: Finding the transportation-land use connection". *Journal of Planning Literature* 13:4, 412–27.

San Joaquin Partnership (1999), *Report to Investors and Members 2000.* (Stockton: SJ Partnership).

Saxenian, A. (1994), *Regional Advantage. Culture and Competition in Silicon Valley and Route 128.* (Cambridge/MA, London/UK: Harvard University Press).

Schamp, E. (1996), "Globalisierung von Produktionsnetzen und Standortsystemen", *Geographische Zeitschrift* 84:3, 205–19.

Schamp, E. (2000), *Vernetzte Produktion. Industriegeographie aus institutioneller Perspektive.* (Darmstadt: Wissenschaftliche Buchgesellschaft).

Schoenberger, E. (2004), "The spatial fix revisited", *Antipode* 36:3, 427–33.

Schönert, M. (2003), "20 Jahre Suburbanisierung der Bevölkerung". *Raumforschung und Raumordnung* 61:6, 457–71.

Schulte, W. (2000), "Die gemeinsame Landesplanung für den Metropolraum Berlin-Brandenburg", *Informationen zur Raumentwicklung* 11/12.1998, 719–26.

Schwartz, B. (ed.) (1976), *The Changing Face of the Suburbs.* (Chicago: The University of Chicago Press).

Scott, M. (1959), *The San Francisco Bay Area. A Metropolis in Perspective.* (Berkeley, Los Angeles: University of California Press).

Senate Administration for Urban Development (2005), Fortschrittsbericht StEP Verkehr. Unpublished Document. Berlin, 10.11.2005.

Senate Administration for Urban Development (2006a), Aktualisierung Stadtentwicklungsplan Verkehr – Mobilitätsprogramm 2011. Unpublished Document. Berlin, 4. June 2006. (Berlin).

Senate Administration for Urban Development (2006b), *Integriertes Wirtschaftsverkehrskonzept.* (Berlin: SenStadt).

Senatsverwaltung für Bauen, Wohnen, Verkehr des Landes Berlin/Ministerium für Stadtentwicklung, Wohnen und Verkehr des Landes Brandenburg (2000), *Integriertes Güterverkehrskonzept Berlin-Brandenburg.* (Berlin, Potsdam).

Senatsverwaltung für Stadtentwicklung (1999), *Planwerk Westraum Berlin. Leitbilder, Konzepte, Strategien.* (Berlin: SenStadt).

Senatsverwaltung für Stadtentwicklung (2000), *Stadtentwicklungsplan Gewerbe. Gewerbestandort Berlin.* Edition Stadtwirtschaft. (Berlin: Regioverlag).

Senatsverwaltung für Stadtentwicklung, Umweltschutz und Technologie; Ministerium für Umwelt, Naturschutz und Raumordnung (ed.) (1998), *Raumordnungsbericht 1998.* (Berlin/Potsdam).

Senatsverwaltung für Stadtentwicklung, Umweltschutz und Technologie (1997), *Gewerbeflächenentwicklung Berlin – Stadträumliches Konzept.* (Berlin: SenStadtUmTech) = Reihe Stadtentwicklung 3.

Senatsverwaltung für Umweltschutz und Stadtentwicklung, Ministerium für Umwelt, Naturschutz und Raumordnung (1995), *Landesentwicklungsplan für den engeren Verflechtungsraum Brandenburg-Berlin (LEP eV).* (Berlin/ Potsdam).

Senatsverwaltung für Wirtschaft und Arbeit (2001), *Wirtschafts- und Arbeitsmarktbericht Berlin.* (Berlin: SenWiArb).

Sheppard, E. (2002), "The Spaces and Times of Globalization: Place, Scale, Networks, and Positionality", *Economic Geography* 78:3, 307–30.

Siedentop, S., Kausch, S., Einig, K. and Gössel, J. (2002), *Siedlungsstrukturelle Veränderungen im Umland der Agglomerationsräume.* Research report to the BBR – First draft. (Dresden: IÖR).

Slack, B. (1998), "Intermodal transportation", in B. Hoyle and R. Knowles (eds) *Modern Transport Geography*, pp. 263–89, 2nd Edn (London: Wiley).

Slack, B. and Frémont, A. (2005), "Transformation of port terminal operations: From the local to the global", *Transport Reviews* 25:1, 117–130.

Slack, B., McCalla, R. and Comtois, C. (2002), Logistics and Maritime Transport: A fundamental transformation. Unpublished paper presented at the 2002 AAG-Conference, Los Angeles/CA, March 2002.

Smith, M. (2001), *Transnational Urbanism. Locating Globalization.* (Malden/ MA, Oxford/UK: Blackwell).

Sonntag, H., Meimbresse, B., Eckstein, W. and Lattner, J. (1998), *Städtischer Wirtschaftsverkehr und logistische Knoten. Wirkungsanalyse von Verknüpfungen der Güterverkehrsnetze auf den städtischen Wirtschafts- und Güterverkehr. Final Report.* (Berlin und Bremen).

SRI International (2002), *Global Impacts of FedEx in The New Economy. Research report.* SRI Center for Science, Technology, and Economic Development (CSTED), Menlo Park/CA. On the web: http://www.sri.com/policy/csted/reports/economics/fedex/ (accessed 1 May 2003).

Stanback, T. (1991), *The New Suburbanization. Challenge to the Central City*. (Boulder et al.: Westview Press).

State of California Employment Development Departement (2007), *Industry Employment for MSAs*, Sacramento (http://www.labormarketinfo.edd.ca.gov/) Accessed on 1 May 2007.

Statistisches Landesamt Berlin (2003), *Aktuelle Daten zur Bevölkerungsentwicklung Berlins*. (Berlin: StaLa).

Stiftung Bauhaus Dessau/RWTH Aachen (1995), *Zukunft aus Amerika. Fordismus in der Zwischenkriegszeit*. (Dessau, Aachen: Stiftung Bauhaus).

Storper, M. (1997), *The Regional World. Territorial Development in a Global Economy*. (New York, London: Guilford).

Storper, M. and Walker, R. (1989), *The Capitalist Imperative. Territory, Technology and Industrial Growth*. (New York, Oxford: Blackwell).

Strauss-Wieder, A. (2001), *Warehousing and Distribution Center Context*. NJPTA Brownfield Economic Redevelopment Project. Prepared for the New Jersey Institute of Technology and the North Jersey Transportation Planning Authority. February 2001. (Westfield: ASW).

Sturgeon, T. (2002), "Modular production networks: a new American model of industrial organization". *Industrial and Corporate Change* 11:3, 451–96.

Suarez-Villa, L. (2003), "The E-economy and the rise of technocapitalism: networks, firms, and transportation", *Growth and Change* 34:4, 390–414.

Sudjic, D. (1993), *The 100 Mile City*. (San Diego, New York, London: Harvest).

Taaffe, E., Gauthier, H. and O'Kelly, M. (1996), *Geography of Transportation*. 2nd Edn (Englewood Cliffs: Prentice Hall).

Taylor, F. (1947), *The Principles of Scientific Management*. (New York: W.W. Norton).

Taylor, G. (1915), *Satellite Cities. A Study of Industrial Suburbs*. (New York, London: Appleton).

Taylor, J. and Jackson, G. (2000), "Conflict, power, and evolution in the intermodal transportation industry's channel of distribution", *Transportation Journal* 39:3, 5–17.

Taylor, M. and Hallsworth, A. (2000), "Power relations and market transformation in the transport sector: the example of the courier services industry", *Journal of Transport Geography* 8:4, 237–47.

Taylor, P., Derudder, B., Saey, P. and Witlox F. (eds) (2007), *Cities in Globalization*. (London and New York: Routledge).

Teaford, J. (1986), *The Twentieth Century American City. Problem, Promise and Reality*. (Baltimore, London: The Johns Hopkins University Press).

Teuteberg, H.-J. (ed.) (1983), *Urbanisierung im 19. und 20. Jahrhundert: Historische und Geographische Aspekte*. (Köln, Wien: Böhlau).

The City of San Leandro (2003), "The San Leandro General Plan 2015". Draft for Public Review. San Leandro: The City of San Leandro (www.ci.san-leandro.ca.us).

The City of Tracy (2001), "Tracy Economic Development Plan". Tracy. (on the web: www.ci.tracy.ca.us).

The Economist, 17th June 2006: The physical Internet. A survey of logistics.

The Lehman-Brothers (1999): Macro shock: How wholesale industries are being revolutionized. Transcript of Lehman-Brothers sponsored Conference Call featuring Adam Fein, Ph.D., June 22, 1999. www.nawpubs.org., accessed 1 May 2001.

Tioga Group (2001), *California Inter-Regional Intermodal System (CIRIS). Rail Shuttle White Paper*. (Moraga: Tioga Group).

Tioga Group in Association with Dowling Associates, Hausrath Economics, TranSystem Corporation (2001), *Port of Oakland – Port Services Location Study*. (Moraga: Tioga Group).

Transportation Research Board (2003), *Integrating Freight Facilities and Operations with Community Goals. A Synthesis of Highway Practice*. (Washington D.C.: TRB) = NCHRP Synthesis 320.

Trip, J. and Bontekoning, Y. (2002), "Integration of Small Freight Flows in the Intermodal Transport System", *Journal of Transport Geography* 10:3, 221–29.

Ullman, E.L. (1980), "Geography as Spatial Interaction", in R. R. Boyce (ed.), *Geography as Spatial Interaction,* pp. 13–27. (Seattle: University of Washington Press).

U.S. Census Bureau (2001), *Population Change and Distribution 1990 to 2000*. Census 2000 Brief, released April 2001. By Perry, M., Mackun, P. et al. (Washington D.C.: U.S. Census Bureau). On the web: factfinder.census.gov, accessed 1 May 2004.

U.S. Census Bureau (2007), *Metropolitan and Micropolitan Statistical Area Estimates* (http://census.gov). Accessed on 1 May 2007.

U.S. Congress, Office of Technology Assessment (1995), *The Technological Reshaping of Metropolitan America*. OTA-ETI-643. (Washington DC: U.S. Government Printing Office).

Urban Land Institute (2004), *Just-in-Time Real Estate. How Trends in Logistics Are Driving Industrial Development*. (Washington D.C.: ULI).

Usbeck, H. (2000), "Aspekte der Suburbanisierung von Gewerbe in ostdeutschen Stadt-Umland-Regionen". *UFZ-Bericht 14/2000*, pp. 21–24. (Leipzig: UFZ).

Van den Berg, L. et al. (1982), *A Study of Growth and Decline*. (Oxford: Pergamon).

Van Klink, H. and van den Berg, G. (1998), "Gateways and intermodalism", *Journal of Transport Geography* 6:1, 11–19.

Van Klink, A. (2002), "The Kempen Nexus. The spatial-economic development of Antwerp and Rotterdam", in Loyen, R., Buyst, D. and Devos, G. (eds) *Struggling for Leadership: Antwerp-Rotterdam Port Competition 1870–2000*, (Heidelberg: Physica), pp. 141–160.

Vance, J. (1964), *Geography and Urban Evolution in the San Francisco Bay Area*. (Berkeley, Los Angeles: University of California Press).

Vance, J.E. (1970), *The Merchant's World. The Geography of Wholesaling.* (Englewood Cliffs: Prentice Hall).

Vigar, G. (2001), *The Politics of Mobility.* (London: Spon Press).

Visser, E.-J. and Lanzendorf, M. (2004), "Mobility and accessibility effects of B2C e-commerce", *Tijdschrift voor Economische en Sociale Geografie* 95:2, 189–205.

Visser, J. and Lambooy, J.-G. (2004), "A dynamic transaction cost perspective on fourth party logistic service development", *Geographische Zeitschrift* 92:1/2, 5–20.

Vogler, G. (1998), "Gewerbeparks im Berliner Umland", in Prigge, W. (ed.): *Peripherie ist überall*, pp. 158–163. (Frankfurt/New York: Campus).

Walker, R. (1981), "A theory of suburbanization: capitalism and the construction of urban spaces in the United States", in Dear, M. and Scott, A. (eds): *Urbanization and urban planning in capitalist society*, 383–429. (London, New York: Methuen).

Walker, R. (1989), "A requiem for corporate geography: new directions in industrial organization, the production of place and uneven development", *Geografiska Annaler* 71B, 43–68.

Walker, R. (2001), "Industry builds the city: the suburbanization of manufacturing in the San Francisco Bay Area, 1850–1940", *Journal of Historical Geography* 27:1, 36–57.

Walker, R. and Lewis, R. (2001), "Beyond the crabgrass frontier: industry and the spread of North American cities, 1850–1950", *Journal of Historical Geography* 27:1, 3–19.

Weber, A. (1909), *Über den Standort der Industrien. Teil 1: Reine Theorie des Standortes.* (Tübingen: Mohr).

Weber, M. (1921), *Wirtschaft und Gesellschaft. Teilband 5: Die Stadt.* Max Weber Gesamtausgabe, edited by W. Nippel. (Tübingen: Mohr Siebeck). Reprint 1999.

Weisbrod, R. (2004), "Ports of the twenty-first century: The age of Aquarius", in Hanley, R. (ed.), *Moving People, Goods, and Information in the 21st Century. The Cutting Edge Infrastructures of Networked Cities.* (New York: Routledge).

Werner, F. (1990), *Ballungsraum Berlin. Raumstrukturen und Planungs-vorstellungen.* (Berlin: TU Berlin) = Beiträge und Materialien zur Regionalen Geographie 4.

Wienert, H. (2002), "Handelsmatrizen als Instrument zur Erfassung von Veränderungen im internationalen Warenhandel", *RWI-Mitteilungen* 53, 149–60.

Wilson, S., and Zambranao, M. (194), "Cocaine, commodity chains, and drug politics: A transnational approach", in Gereffi, G. and Korzeniewicz, M. (eds), *Commodity Chains and Global Capitalism,* pp. 297–315 (Westport: Praeger).

Womack, J., Jones, D. and Roos, D. (1990): *The Machine that Changed the World.* (New York: Rawson).

Woudsma, C. (1999), "NAFTA and Canada-US cross-border freight transportation", *Journal of Transport Geography* 7:2, 105–19.

Woudsma, C. (2001), "Understanding the movement of goods, not people: issues, evidence and potential", *Urban Studies* 38:13, 2439–5.

Woxenius, J. (2002), "Conceptual Modelling of an Intermodal Express Transport System", International Congress on Freight Transport Automation and Multimodality Delft, 23–24 May 2002. (Delft: FTAM).

Wrigley, N. (2000), "The globalization of retail capital: Themes for economic geography", in Clark, G., Feldman, M. and Gertler, M. (eds) *The Oxford Handbook of Economic Geography*, pp. 293–313. (Oxford: Oxford University Press).

Wrigley, N. and M. Lowe (eds) (1996), *Retailing, Consumption and Capital. Towards The New Retail Geography* (University of Southampton: Longman Group).

WTO (2002), International Trade Statistics, http://www.wto.org/english/res_e/statis_e/statis_e.htm. (accessed on 1 May 2003).

Zimm, A. (1959), *Die Entwicklung des Industriestandortes Berlin. Tendenzen der geographischen Lokalisation bei den Berliner Industriezweigen von überörtlicher Bedeutung sowie die territoriale Stadtentwicklung.* (Berlin: VEB Deutscher Verlag der Wissenschaften).

Zimm, A. (1991), "Raum-zeitliche Etappen der Metropolenbildung Berlins, ein Blick zurück nach vorn", *Petermanns Geographische Mitteilungen* 135, 99–111.

Zimm, A. (ed.) (1988), *Berlin und sein Umland. Eine geographische Monographie.* (Gotha: VEB Hermann Haack Geographisch-Kartographische Anstalt).

ZLU/Zentrum für Logistik und Unternehmensplanung (2001), Güterverkehrsaufkommen. Daten zum Güterverkehrskonzept Berlin-Brandenburg. Unpublished manuscript. (Berlin: ZLU).

Index